T0309514

The central models of theoretical physics have been extraordinarily successful in describing and predicting the behaviour of physical systems under an enormous range of conditions. But why are these mathematical theories so successful, and how is their structure influenced by the nature of the observations on which they are inevitably based? This intriguing book examines these subtle and fundamental issues, and investigates the complex interdependency of theory and experiment.

The dependence of the theories of quantum mechanics and relativity upon measurements and standards, dealt with in the opening chapters, leads to a discussion of the uncertainties inherent in the physics of systems displaying chaotic dynamics. The reasons why mathematical theories of physics are effective are then discussed, and related to constraints on observation. The book concludes by arguing that successful prediction provides compelling support for belief in a world independent of the observer.

Dealing with important and basic aspects of the general framework of physics, this book will appeal to undergraduate and graduate students in the physical sciences, and to anyone with an interest in the empirical and metaphysical foundations of science.

The Observational Foundations of Physics

The Observational Foundations of Physics

Sir Alan Cook
formerly Master of Selwyn College, Cambridge

CAMBRIDGE UNIVERSITY PRESS
Cambridge, New York, Melbourne, Madrid, Cape Town, Singapore,
São Paulo, Delhi, Dubai, Tokyo

Cambridge University Press
The Edinburgh Building, Cambridge CB2 8RU, UK

Published in the United States of America by Cambridge University Press, New York

www.cambridge.org
Information on this title: www.cambridge.org/9780521454506

First published 1994

A catalogue record for this publication is available from the British Library

Library of Congress Cataloguing in Publication data

Cook, Alan H.
The observational foundations of physics/Sir Alan Cook.
p. cm.
Includes bibliographical references.
ISBN 0-521-45450-6. – ISBN 0-521-45597-9 (pbk.)
1. Mathematical physics. 2. Physical measurements.
QC20.C684 1994
530′.01 – dc20 93-36901 CIP

ISBN 978-0-521-45450-6 Hardback
ISBN 978-0-521-45597-8 Paperback

Transferred to digital printing 2010

CONTENTS

PREFACE x

1 Introduction 1

1.1 The questions 1
1.2 The nature of observation and of theory 2
1.3 Measurement and standards 7
1.4 The standard of frequency and standards derived from it 9
1.5 Theory in an uncertain world 12
1.6 Beyond physics 14

2 Standards of time and equations of motion 16

2.1 Introduction – a question of tautology? 16
2.2 An operational analysis 19
2.3 Classical standards of time and equations of motion 24
2.4 Defining constants of physics 26
2.5 Summary 28

3 Observations at a distance: special relativity 30

3.1 Introduction 30
3.2 Geometrical observations at a distance 34

3.3 4-vectors as representing physical observations 41
3.4 Electromagnetism 46
3.5 General relativity 49
3.6 Conclusion 51

4 Microphysics: relativistic quantum mechanics 52

4.1 Introduction 52
4.2 Quantum mechanics in the geometry of special relativity 54
4.3 From observation to theory 59
4.4 Conclusion 67

5 Indeterminacy in theory and observation 69

5.1 Introduction 69
5.2 Chaos 72
5.3 Observables and predictions of chaotic dynamics 79
5.4 Systems of many bodies 81
5.5 Nearly soluble problems of many bodies 85
5.6 Thermodynamics and statistical mechanics 90
5.7 Observables of many-body dynamics 93
5.8 Summary and conclusion 94

6 Why does mathematical physics work? 96

6.1 Introduction: the problem 96
6.2 Abstract groups and concrete realisations 101
6.3 Physical representations of groups 105
6.4 Symmetry: finite groups 108
6.5 Continuous groups: metrology 109
6.6 Dynamics 110
6.7 Physics and group representations 113
6.8 Dynamics and semi-groups 117
6.9 Conclusion 119

7 Probable argument 121

7.1 Introduction 121
7.2 Basis of probable argument 123
7.3 Mathematical probability 126

7.4 Scientific inference 130
7.5 Sources of uncertainty 133
7.6 Predictability and its implications 135

8 Conclusion 140

8.1 Introduction: the content of observation 140
8.2 Measurement, observation and theory 142
8.3 The indefiniteness of nature 145
8.4 The coherence of nature 146
8.5 Back to a real world? 148

APPENDIX 153

REFERENCES 156

INDEX 160

PREFACE

PHYSICS AS A RATIONAL humane study of the world around us has three particularly characteristic features: it relies heavily on measurement for observing the nature and behaviour of the world, it has clear powerful logical structures and it is able to make effective predictions of phenomena still to be observed. It is however plain that much of the knowledge which constitutes physics is to some degree uncertain. I try in this book to show how those aspects of physics are related.

I argue in the first place that the forms of certain theoretical structures are determined by the nature and limitations of the ways in which we are able to make measurements. I extend the argument to physical processes of which the outcome is inherently uncertain and consider what it may be possible to measure about such processes. One of the great puzzles of theoretical physics that has often exercised people is why mathematics is so effective in theoretical physics, and I put forward an explanation based on correspondences between natural processes and representations of abstract mathematical groups. Ideal models however never correspond exactly to nature; the discrepancies lead us to consider the place of probable argument in physics.

Throughout the argument two questions persist in the background and come forward from time to time. The first is, why observations of the natural world seem to fit so well into strict logical structures – is it because the world really is like that or is it because of the way in which we select for study those phenomena that fit into such structures? The second question is why, in the face of uncertain knowledge and of questions about the rational status of the natural world, physics is so good at prediction? Those questions are the subjects of the concluding chapter.

The matters that I attempt to illuminate in this book are far from novel but two developments in recent years have made it timely to raise them again. One is the strictly technical metrological decision to adopt an atomic standard as the fundamental standard of frequency and to derive from it all other standards of measurement (except that of mass) through certain quantum processes and the definition of related defining constants such as the speed of light. That development was undertaken for purely technical advantage, but the consequence has been to clarify the status of the time evolution equation in quantum mechanics and of the transformations of special relativity.

The other development has been the recognition of the importance of non-linear phenomena in physics and of chaotic dynamics. As a result it is now more widely appreciated that some phenomena in physics are inherently uncertain, not just that our observations are inadequate, and that sharpens up the discussion of what is or is not predictable in physics. It also bears on the question of the place of mathematics in theory, for it seems that a different type of mathematics is appropriate to chaotic phenomena, and that in turn helps our understanding of why certain branches of mathematics are so effective in more traditional physics.

This book is about physics and the particular arguments in Chapters 2, 3 and 4 are about physics alone. Parts of the argument however have a much wider application. I consider that science is far more than pattern recognition. It is not sufficient to say that nature is as it is because it is as it is. Science sets out to account for the natural world in a logical way as the consequence of a few premisses. We have to see the reason behind the pattern. Logical structures are just as important in geology or biology as in physics. Questions of probable argument and prediction are also ubiquitous, while an important development in recent years has been the application of dynamical arguments to biological processes. This then is a book for physicists, but some of the most profound methods and problems of physics are not confined to physics and so parts of the book may be found useful or stimulating to anyone who wants to come to a rational understanding of the natural world and how we study it.

I have discussed the topics of the book with a number of people over the years, often in quite an informal way so that there are few whom I can single out specifically, but some I must mention. I am greatly indebted to Dr T. J. Quinn and Dr B. W. Petley for helping to form my views on metrology and the place of standards of measurement. One of the anonymous readers of my first draft raised a number of philosophical issues and encouraged me to adopt a more philosophical approach than I had done originally and Professor Onora O'Neill has provided valuable criticism of some of my philosophical propositions. Lastly, I am indebted to Dr S. Capelin of the Cambridge University Press for his editorial support.

A.H.C.

1

Introduction

1.1 The questions

MY PURPOSE IN THIS BOOK is practical and empirical, it is to attempt to unravel some ways in which the practice of physics determines the form and content of physics and physical theory. That is no novel undertaking. Eddington (1953) in particular claimed to derive many fundamental features of physics from deep epistemological principles, but as is well known, few have understood what he was about and fewer still have agreed that he was successful. My aim is less ambitious than that or of some philosophical discussions, it is to look at what physicists actually do in making observations and assessing their reliability, and to follow through the consequences of those practices for the theoretical structures of physics. Modest though that may seem, we shall find that it leads us into quite deep and intractable questions concerning the status of observation, the basis of inference and the reliability of physical knowledge.

A very striking feature of the physical sciences is that they are remarkably effective at predicting from past phenomena the nature of events yet to take place. Why should physics be so effective, and what does that tell us about the world of physics and our ways of gaining knowledge of it? The question has become the more acute as it is realised that much of the behaviour of the natural world is at bottom chaotic, in the sense that conditions cannot be stated precisely enough for consequences to be predicted.

Most of this book is about physics in a rather restricted way and only at the end do I take up epistemological questions such as those at which I have just hinted, but there is one philosophical issue that must be faced at the beginning and then put aside until the last chapter. That is the question of the existence of a physical world independent of us, or more strictly, independent of me. Is there a real world that exists independently of whether I or anyone else is looking at it, or are all the ideas I have about a world external to me just the construction of my mind? My own opinion is that no answer can be given to that question. Either the realist position or the extreme idealist position (solipsism) can be the basis for a consistent account of what goes on in my mind, although I consider that it is difficult to hold any consistent intermediate position. In this book I write as a realist. In the first place, it is far more straightforward to do so than to write consistently as a solipsist. More importantly, and in the spirit of the overall approach I adopt, I believe that almost all physicists when working at the bench or with their pencil and paper or computers, behave as if there were a real world that will continue to exist whether or not they observe it or think about it (see d'Espagnat, 1989). Thus Pickering (1989) and Gooding, Pinch and Schaffer (1989) in their respective discussions of the *Uses of Experiment* explicitly accept that a real material world exists, and most of this book is written unquestioningly in that belief, but I shall return to it with other metaphysical matters at the end.

1.2 The nature of observation and of theory

While most physicists, so I think, pursue their vocation accepting the existence of a real world, independent of themselves, out there to be investigated, few are so naïve as to think that their observations give them direct unadulterated knowledge of that world. The formalism of quantum mechanics expresses the idea that our observations are the results of interactions between the world independent of us and the process of observing it, and the development of quantum mechanics has led scientists in other fields also to appreciate that the results of experiment and observation depend on how we interact with the outside world when we perform those experiments and observations. I should comment here that I make no distinction between *experiment* and *observation*. Experiment commonly implies a more active approach to nature on the part of the observer, while observation is usually considered to be more passive. Those distinctions are irrelevant to the argument of this book.

There are two aspects to that dependence of the results of observation on our interaction with the outside world, an objective aspect and a subjective one, or, to put it slightly differently, a dependence that is the consequence of the physics and independent of the observer, and a dependence that follows from the personal competence or choices of the observer. The dependence incorporated in quantum mechanics is objective – it is expressed by operators of definite mathematical form, human factors do not come in. The dependence that comes from the design of experiment or technical competence is peculiar to the people doing the experiments. Social influences come in here. Without going all the way with sociologists of science who sometimes seem to imply that our view of nature has nothing objective to it at all but is entirely a social construct, or with literary theorists who would have us believe there is nothing beyond a text, it is still possible to recognise that the subjects on which physicists work and the ways in which they approach them, are certainly influenced by communal behaviour, although not wholly determined by it. Ziman (1978) asserted that the claim of science to be objective depends upon its being a social construct, created cooperatively, it is in his words, in the noetic domain, and d'Espagnat (1989) has made an important distinction between the subjective knowledge of an individual and the subjective elements in knowledge, common to a large group of individuals, on which as scientists they all agree. I am concerned in this book with that communally accepted body of knowledge.

Physics is supposed to be empirical and contingent, with observation primary and theory secondary but many philosophers of science have questioned that position. Hesse (1974) argued in some detail that even apparently simple observations depend on some theory for the interpretation of the raw response of an instrument, and her argument could be put more forcibly today (Hacking, 1983). Consider for example, the measurement of the intensity of light emitted by a black body. The light falls on a semiconductor detector that generates an electrical signal that causes another semiconductor device, a digital voltmeter, to emit a train of electrical pulses that set up a certain state in the memory circuits (more semiconductors) of a digital computer. The computer also receives a train of signals from a second complex of electronic devices that purports to measure the temperature of the black body. Finally the computer, having itself issued electronic instructions for changing the temperature, calculates a relation between temperature and light intensity. It would be difficult to argue that the result is independent of theory (see also the similar analysis in Toraldo di Francia, 1981). Hesse's argument might seem to apply equally forcibly to the realisation of the standard of frequency which is described below in this chapter and the next.

All observation or experiment involves three elements, a purpose which arises from some prior theoretical issue, physical instruments and procedures, and an abstract model of those instruments and procedures with which the 'result' of the observation or experiment is calculated. Hesse's argument is that instruments and procedures themselves as well as the interpretation of their readings, depend on pre-existing theory, and that can hardly be gainsaid. Furthermore, as Pickering (1989) has insisted, observations as the outcome of experiments incorporate available instrumental techniques as well as the theoretical underpinning.

When the 'result' of an experiment is finally established it may or may not agree with the theoretical scheme which prompted the observation in the first place. One obvious reason for that is that the scheme is an inadequate representation of nature, that the observation encounters what Pickering (1989) calls 'resistances'. In that case the observations provide new knowledge about the natural world. The 'result' may also disagree with the initial theoretical scheme because the model used to derive it from the raw data does not correspond closely enough with the physical processes and relations. Observations are never interpreted independently of some abstract model of the physical system, analyses and calculations of results are done on the model quantities which are supposed to correspond to the physical quantities. Thus we make some calculations with a 'voltage'. We do not observe a 'voltage' directly, but rather some symbols on a digital voltmeter that are intended to correspond to the abstract notion of 'voltage' as established by the theory of the instrument and the manner of its construction.

Even in the very simplest cases, abstract representations of the physical state are involved, as in the calculation of the volume of a nominally regular solid from measurements of its dimensions (see Cook, 1961, 1975). The model used is the relation between volume, V, and the position vector, $\mathbf{r}$ of an element $\mathrm{d}S$ of the bounding surface:

$$\int_S \mathbf{r}\cdot\mathrm{d}S = \int_V \mathrm{div}\,\mathbf{r}\,\mathrm{d}V = 3\int_V \mathrm{d}V,$$

or

$$V = \frac{1}{3}\int_S \mathbf{r}\cdot\mathrm{d}S.$$

That simple result can be deceptive: it assumes that we know the position vector for each surface element, measured from the same origin, but since it is usual to measure distances, $\Delta\mathbf{r}$, *between* surface elements, for example across diameters of a supposed sphere, hidden shifts of origin can occur leading to erroneous calculations. Those and other possible discordances are not negligible if a volume is to be determined to one part in ten million. The discrepancies that arise as a result of an inadequate model of an experiment or observation are what we call systematic errors, and it is well known that they can be very difficult to identify.

Analyses of the reasons for making observations, of the ways in which they are made, of the theory and technology on which they depend, are no doubt of considerable interest, but they do not of themselves invalidate the results of observation or theory. Indeed it can be argued that far from casting doubt on an experimental result, the fact that a procedure is based on well established theory gives an assurance that the result is telling us something about the real natural world (see Franklin, 1989).

I assert, however, that analyses of that sort are irrelevant to the argument that I shall develop in this book. I take the results of observation as a physicist presents them, for however they were obtained they are the empirical basis of physics, or as Toraldo di

Francia (1981) puts it, *a physical quantity is defined by prescribing the operations that are carried out in order to measure it*. I seek to understand how the results of the actual practice of observation, the data as they are, determine the structure of theory. Theory in the first place must bring order into the results of observations that physicists carry out and my main purpose is to see how the one determines the other. Whether or not theory then tells us something about the real world behind the observations is a question that I defer to the final chapter.

I adopt in this book the concept of a theory as a model of our observations of the real world – not a model of that real world itself which, as I have asserted, is not directly accessible to us, but explicitly a model of the results of observations of the world, with observations defined by the operations which produce them as I have just explained. There is a view of theory, the 'instrumentalist' which maintains that a theory is just a means, an instrument, a way of calculating the outcome of observations, and that the content and structure of the theory do not necessarily bear any relation to the independent world behind the observations. An operational view of observation and an instrumentalist account of theory, while they are evidently consistent, do not necessarily entail the one the other. I do adopt throughout the operational account of observation but I consider that a true theory is more than just a calculating machine. I take a theory to be a mathematical realisation of an abstract system that has properties corresponding to those of a set of observations, a concept which I shall develop in subsequent chapters and that is at the heart of much of my argument. It is in that sense that I take a theory to be a model of the world of observations, with the implication that there is a more fundamental correspondence than just giving the right answers, and my aim in this book is to show how that fundamental correspondence comes about. That view of theory might superficially seem to be similar to Plato's notion of the relation of our world of sense impressions to an ideal mathematical world. There is, however, a deep distinction, for Plato considered the physical world that we experience to be an imperfect realisation of the more real formal structures of the mathematical world, whereas my position is that the physical world is primary and the abstract system is the best we can do to represent it.

1.3 Measurement and standards

My concern in this book is with the objective factors in the interaction between nature and the observer, of which one of the most important is the process of measurement (see Cook, 1977, 1992). Measurement is the basis of all physical science and the consequences of the constraints that it imposes are the topics of the next two chapters. I therefore go on to summarise the nature of measurement, to describe the system of standards on which physical measurements are based, and to relate them to the equations of physics.

All equations of physics, for example, Newton's equation of motion, $\partial \mathbf{v}/\partial t = \mathbf{F}/m$, are representations of physical states or processes in that they are mathematical relations that are congruent to the relations between observations. Measurements are necessary to establish the correspondences and to ensure that the representations of reputedly similar observations are compatible.

Every measurement consists of comparing some quantity with a standard quantity of the same type, and thus assigning a number to the measure of the unknown quantity in terms of the standard. Lengths are measured by setting objects alongside other objects on which standard lengths are marked out. Times and frequencies are measured by comparing them with times or frequencies of electrical signals derived from some standard oscillator. Masses are measured by comparing them on a balance with standard masses. Electrical voltages and currents are measured by setting them directly against voltages and currents supplied by standard sources. The precision of any measurement is determined both by the accuracy of the comparison with the standard and by the precision with which the standard quantity can be realised and reproduced. The system of standards of physical quantities affects all physics and all applications in engineering, and it will be argued in the next two chapters that the ways in which we measure and the choice of basic standards determine also some of the fundamental structures of physics. The nature of the system of basic standards is crucial to the argument and to it I now turn.

It is well known that there is no need to have a separate standard for every physical quantity and that in fact standards for all physical measurements can be derived from just four independent standards, conventionally those for time, length, mass and electrical current. The Système International des Unités (SI) has as its fundamental independent units the Second, the Metre, the Kilogramme and the Ampère, but that statement is already somewhat out of date, if not misleading, for it implies that there are indeed distinct physical objects, such as the standard metre, by which those standard units are realised. In fact the standard of length is derived from that of frequency by an independent value of a fundamental constant, the speed of light. The unit of length is the distance travelled by electromagnetic radiation in free space in a specified time, and its value in terms of the conventional metre is derived from the standard of time and a conventional value (2.997 924 58 m/s) adopted internationally for the speed of light (Appendix – Resolution A4 of the XXI General Assembly of the International Astronomical Union, 1991). The particular numerical value ensures that lengths derived with it are consistent with those derived from earlier physical standards, but the precision with which lengths can be derived from light times is greater than that of realisation of the now superseded physical standards of length.

Similarly the standard of electrical voltage can be derived from the standard of frequency through the Josephson effect and a conventional value of the ratio h/e of Planck's constant to the electronic charge. Here again the precision with which a voltage can be derived in that way is better than the accuracy of the value of the ratio h/e in the terms of the electrodynamical standards of electrical units. The standard of mass remains as yet unrelated to the standard of frequency but since it is now possible to relate the unit of electrical current to that of voltage through the resistance of the quantum Hall effect, it is conceivable that the unit of mass could be replaced by a unit of energy derived from the units of electrical current and voltage and hence related to the standard of frequency through the two constants of the Josephson effect and the quantum Hall effect.

Nowadays then, all the other fundamental units can be related to the standard of frequency through conventional values of certain constants of physics. It should be appreciated that relating other quantities to frequency has not reduced the number of independent quantities on which the system of measurement is based. Although we no longer use an independent metal bar to realise the standard of length, the conventional value of the velocity of light that we use is equally arbitrary and is as much an independent physical element of the system of standards as the metal bar.

The standard of frequency is itself realised by an atomic process and two of the three constants arise from quantum processes, the Josephson effect and the quantum Hall effect. Quantum physics has thus to a large extent replaced classical physics as the basis of the standards of measurement (Petley, 1985). The reason for that is nothing very subtle, it is simply that the resulting system of standards, units and measurement, depending as it does on various electrical and electronic measurements, is more convenient in use, more precise and more generally accessible, than the mechanical system it has replaced. At the same time, the implications for the logical structure of physics are profound, as will be argued in the next two chapters; or rather, the use of the new scheme of units and standards reveals that structure more clearly than may often have appeared in the past.

1.4 The standard of frequency and standards derived from it

Before drawing out the implications of the new system, the physical nature of the standard of frequency and of the constants of physics, as well as their logical position in the scheme of physics, must be explained. The present standard of frequency is the frequency of an electrical signal that causes transitions between the two hyperfine levels in the ground state of the atom caesium-133 (Appendix – Resolution I of the XIII CGPM, 1967). The standard is realised physically in apparatus in which the atoms in an atomic beam of caesium-133 first pass through a magnetic filter that prepares them in their hyperfine states as distinguished by their magnetisation. They then pass through a region in which an electromagnetic field is maintained at the correct frequency (9192.6 MHz) followed by a second magnetic filter to detect when transitions between the hyperfine levels have occurred. It is found in practice that the frequencies of the electrical signals realised in that way in different laboratories agree to within about 1 part in

10^{13} or better. The standard is therefore highly reproducible; it is also widely and easily accessible through radio transmissions. A somewhat more convenient apparatus is the atomic hydrogen maser, in which an inversion of the populations of the two hyperfine levels in the ground state of atomic hydrogen is brought about by a magnetic filter. Stimulated emission from the upper level maintains electrical oscillations in a cavity tuned to the microwave frequency of the transition, about 1420 MHz. The maser is convenient because it generates a continuous electrical signal, but the frequency depends to some extent on coupling with the microwave resonant cavity and so is considered to be less fundamental than that of the caesium standard (Kartaschoff, 1978).

It is evident that both forms of standard depend heavily for their design and operation on theory, not only for the basic principle, but in the operations of the source, the detector and the filter. None of that invalidates the status of the apparatus as a means of realising a fundamental standard, for by an internationally agreed definition, the standard frequency is the frequency generated or identified in the operation of that apparatus. All that is necessary is that design, construction and operation of the apparatus should be so closely specified that everyone who operates an example of it to the specification should get consistent results.

Once a standard has been defined for some quantity, it is meaningless to speak of how it may change or of checking it against some other standard. The second was originally defined by the rate of rotation of the Earth upon its axis and when it was suspected that the Earth's rate of rotation might vary, the second (now the ephemeris second) was re-defined in terms of the period of the Earth in its orbit about the Sun. It then became meaningful to talk about the variable spin of the Earth whereas previously it had not because there had been no better standard against which to test the spin. We have to recognise, as Wittgenstein has emphasised, that some apparently well formulated questions have no answers. Now that the second is defined by an atomic process, we can in turn discuss the possible variation of the ephemeris second, but it is meaningless to speak of changes in the atomic standard itself unless an improved way of defining the second is developed and replaces the caesium standard by general consent. There is of course a practical question of how to define the standard frequency when apparatus in different laboratories may give slightly different results, but that is done by international agreement on a day-to-day basis and will give a figure that is within about 1 part in 10^{13} of the nominal value. The reproducibility of comparisons is now so good that it is necessary to take into account general relativistic shifts of frequency when apparatus is in different gravitational fields (Appendix – Resolution A4 of the XXI General Assembly of the International Astronomical Union, 1991). By definition the value of the standard cannot change with time and so the standard frequency is to be regarded as represented by an electrical signal with a single Fourier component to within 1 part in 10^{13}. The standard of frequency is thus defined operationally, by pointing to a specific type of apparatus and asserting that it produces an electrical signal with a defined frequency adopted as a standard and therefore invariant: Chapter 2 is concerned with the implications for the structure of quantum physics.

The constants that are used to derive other standards from the standard of frequency are also defined operationally and are likewise invariant. We agree internationally that the speed of light shall be taken to be 2.997 924 58 m/s (Appendix – Recommendation II of the General Assembly of the International Astronomical Union, 1991) and that all lengths shall be derived from the time that electrical signals take to traverse them at that speed. The independent standard of length is abandoned and there is now no way of checking whether the speed of light is a constant. It is by its definition a constant. So similarly are the factors by which electrical current and voltage are related to frequency. The reasons for changing from an independent standard of length to one related to frequency are that the measurements themselves are more precise, and that the definition of length conforms to the fact that all distances beyond a few metres have in practice to be measured by electromagnetic signalling rather than by direct comparisons with standard bars. The implications of that state of affairs for the local geometry of space-time, namely special relativity, are developed in Chapter 3, along with the further implications of the choice of electrical standards for electrodynamics and relativistic dynamics. Special relativity applies only in empty space and observations show that close to massive bodies it must be extended as general relativity.

It is natural at this point to comment on the justification for using the space-time geometry of special relativity in microscopic physics as well as on the large scale; that is the subject of Chapter 4, where questions raised in the two preceding chapters are raised again but from rather different points of view.

1.5 Theory in an uncertain world

The theories discussed in Chapters 2, 3 and 4 are linear theories, the operators in the mathematics are linear operators. They are strictly linear because they incorporate linear transformations of coordinates. In much of physics however, the interactions between physical objects are non-linear to a greater or less degree and that may give rise to behaviour that is called chaotic. In consequence there are aspects of physics that are inherently uncertain and it is not possible to make unambiguous observations. Another important cause of uncertainty lies in the fact that the equations of motion of systems of three or more bodies have no completely definite solutions. Here the problem lies more in the formulation and manipulation of the mathematical models rather than in the dynamics, for definitive physical systems of three or more bodies certainly exist. Chaotic dynamics and systems of many bodies, discussed in Chapter 5, pose questions of what are meaningful observations of erratic physics.

Mathematical arguments are used so ubiquitously in physics that they seem essential to it, and while it may be that all or most of the structure of physics could be expressed in words, it is certain that it would be exceedingly cumbersome and inefficient to do so and the arguments would be very much less direct and clear. In addition, it would be difficult to express the results of measurements so effectively or to analyse them so powerfully, without doing so numerically. Why is it that mathematics appears as almost essential to physics, is it because the world is made that way, a notion that goes back to the Pythagoreans (Pedersen, 1993) or is it because we choose to study those aspects of the world that can be put into mathematical form, as Ziman (1978) considered, or do we bend the world to make it conform to our mathematics? Many physicists and philosophers of physics have confessed to being puzzled by those questions. I take up the matter in Chapter 6, beginning with the notion that the mathematical forms of theories of physics are in some sense a consequence of the restrictions imposed upon our physical knowledge by the ways in which we measure. It might appear then, that the whole structure of physical theory is the outcome of the ways in which we observe and measure, something imposed on our knowledge of that real world independent of us; yet we may also wonder whether the possibilities of measurement that are open to us are not themselves dependent on the nature of the real world – if for example, there were no long-lived hyperfine states in atoms, could we establish standards of frequency in the way we do?

It is a striking feature of physical science that, more than any other human study, it can predict the results of experiments or observations not yet performed. All successful theories were developed to bring order into sets of empirical observations and in that sense are examples of inductive reasoning, from empirical instances to general principles. When however, those general principles seemed reasonably secure, they would become the basis of deductive reasoning and give predictions of possible observations not yet made. Often, but not always, predictions of physical theories turn out to agree well with observation. Why should that be? We are again met with the question, is that the way the world is, really rational and predictable, or have we selected in some way those features of it that are necessarily reproducible? The study of chaotic behaviour has forced us to realise that many aspects of the natural world cannot, for quite fundamental reasons, be predicted in detail, yet despite that, prediction is frequently very successful. The answers to those conundrums must in part concern the scope of a theory, they must also involve the reliability of arguments by which we arrive at general principles from empirical observation.

Elucidating the ways in which theories in physics arise and are accepted is an ongoing major preoccupation of philosophers and historians of the sciences. It is very rare that any major new insight into the natural world has come about inductively by the Baconian method of assembling large sets of data and deriving general laws from them. Far more typically a flash of imagination based on very few observations, leads to a theoretical structure, usually in the form of a mathematical system by which the results of further observations can be calculated. Those calculations are then compared with observation and if there is satisfactory agreement, the theoretical model is accepted as a means of predicting yet other observations.

1.6 Beyond physics

That summary description glides over some profound questions – what do we mean by 'satisfactory agreement'? and supposing the further observations are not quite in agreement with predictions, is that reason for rejecting the theory or may it be just what we might have expected in an uncertain world? Those are questions about probabilities and probable argument and probability is the subject of Chapter 7. The meaning of probability and the nature of probable argument have been the concern of philosophers, natural philosophers and mathematicians since early in the sixteenth century, but the last century has seen particularly lively and wide ranging debates on these matters, due in large part to the use of statistical methods to study the results of experiments in the biological sciences as well as, more recently, in economics and social studies. Physical scientists have on the whole not engaged much in those arguments, and have often adopted the view that if you had to use statistical methods to derive a result from some experiment, then you had better devise a better experiment. There is force in that, but it is not the whole story, and cannot apply to those sciences, such as astrophysics and geophysics, that involve observations of objects that we have to take as given. Probable argument and the conditions it imposes on the structure of physics lie behind the formidable face of physical science and must be considered. They are the subject of Chapter 7 and bring us back to the social or communal nature of science through such questions as, if I assert that some statement in physics has a certain probability, is that an assertion which is no more than my own, or can that probability in principle have an objective value of which the value that I assign is an estimate? Between those two views, the purely subjective and the formally objective, is there a third, in which probabilities are assessed by communal agreement of a social group? Another issue is the relation of probabilities as bases for action to probabilities as grounds for accepting some theory as a 'true' representation of the physical world, and that in turn seems to lead back to prediction in physics and what we can infer from the frequent successes of prediction (Jeffreys and Wrinch, 1921).

I have written of the physical sciences so far, and most of the book is about physics, but some of the questions arise also in the biological sciences. What are the right abstract models for biological processes and states? what is the place of probable argument? does prediction have the same status and force as in physics? I shall say a little on those matters at the end. I also return to more general issues of philosophy. The main philosophical positions I have adopted are that there is a world independent of me, that I and other scientists learn about that world through observations that we all accept, that we define the results of those observations operationally, and that theories are in the first instance means ('instruments') for calculating the results of observations. I then show how the nature of observations leads to the structure of theories. But are our theories solely instruments or do they also tell us significant things about the world independent of me? How much can science tell us about an external world, supposing that to exist? I argue that a successful theory is in fact more than a calculating instrument and the fact that it is successful means that it corresponds in some way to objective features of our observations of the natural world.

2

Standards of time and equations of motion

2.1 Introduction – a question of tautology?

THE STANDARD OF FREQUENCY, as was seen in Chapter 1, is the frequency of electromagnetic radiation that causes transitions between two particular hyperfine levels in the ground state of Cs-133. Caesium is an alkali metal and thus in the ground state, with principal quantum number 1, it has the electronic configuration:

$$\mathcal{J} = \tfrac{1}{2}, \quad L = 0, \quad S = \tfrac{1}{2},$$

in the Russell–Saunders (L–S) coupling scheme.

The caesium nuclide of mass 133 has spin I of 7/2 and so the possible values of the total spin, F, equal to $(I \pm \mathcal{J})$, are 4 and 3; in the former the electronic and nuclear spins are parallel, in the latter anti-parallel.

The possible values of m_F, the projection of F on the direction of an external magnetic field (Figure 2.1), range from -4 to $+4$ for $F = 4$, and from -3 to $+3$ for $F = 3$. If m_F is zero and remains zero in a transition from $F = 3$ to $F = 4$, the effective magnetic dipole moment of an atom of Cs-133 changes sign in the transition, and so when an atom with $F = 4$ passes through a region of non-uniform magnetic field, it experiences a force in the opposite sense to that experienced by an atom with $F = 3$. It is therefore deflected

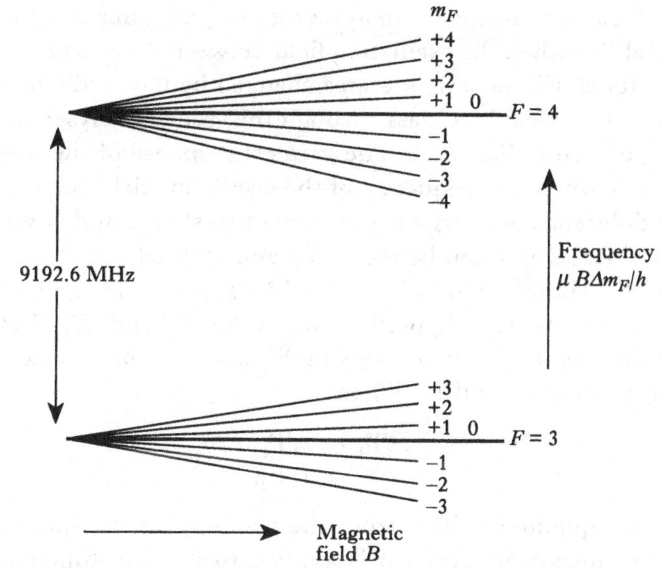

Figure 2.1. Zeeman diagram of ground state of Cs-133.

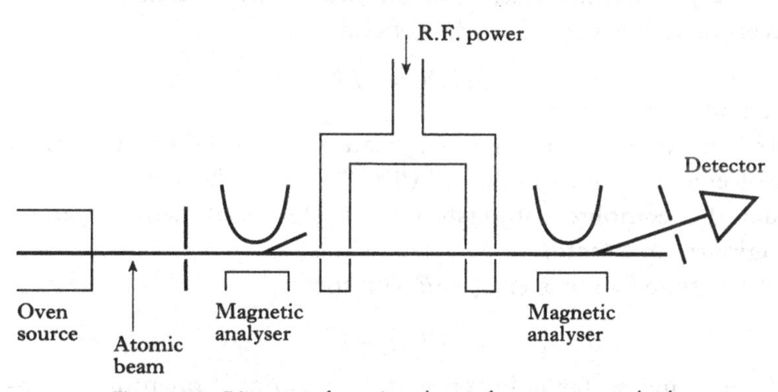

Figure 2.2. Diagram of caesium beam frequency standard.

in the opposite direction. If then a beam of caesium atoms is passed through a region with a non-uniform magnetic field, it can be separated into two beams of atoms with different values of F; such an arrangement is a magnetic analyser.

Transitions between the one hyperfine state and the other can be induced by an electromagnetic field at the frequency corresponding to the difference in energy of the two states (9192.6 MHz). Suppose that a beam of atoms passes through a magnetic analyser, then through a region with an oscillating electromagnetic field and lastly through a second magnetic analyser set to reject atoms that pass the first analyser when the oscillating field between them is turned off. Some atoms will have their states changed by the oscillating field and will then be able to pass through the second analyser and fall upon a detector (Fig. 2.2). The greatest response of the detector will occur when the frequency of the oscillating field corresponds to the difference of energy between the two states; in that way the standard frequency can be identified and realised.

Let the Hamiltonian of the Cs-133 atom be H, and let the energies of the two hyperfine states be E_1 and E_2. Let the wavefunctions of the two states be Ψ_1 and Ψ_2, corresponding to the eigenvalues E_1 and E_2. Then

$$H\Psi_1 = E_1\Psi_1,$$
$$H\Psi_2 = E_2\Psi_2.$$

Those equations follow from the time-evolution equation for quantum mechanics when it is applied to the wavefunction of a stationary state, that is, one for which the intensity of the wavefunction is a constant. In general

$$\mathrm{i}\hbar\partial_t\Psi = H\Psi$$

(∂_t stands for $\partial/\partial t$).

If Ψ corresponds to a stationary state, it is one of the orthonormal set of eigenfunctions of H and $(\Psi^* \mid \Psi) = 1$. (The star denotes as usual the complex conjugate.) Ψ is Hermitian with a purely imaginary time factor.

If the time factor is $\exp(-\mathrm{i}Et/\hbar)$, then

$$\mathrm{i}\hbar\partial_t\Psi = E\Psi.$$

Everything so far is in terms of conventional quantum mechanics, which at once raises a question – it seems that a standard of time has already been introduced through the operator ∂_t in the fundamental equation (and also through the definitions of E and $\hbar$) and that now, apparently, we are going to re-define the standard of time (that is, frequency) through a specific experimental operation based upon those same equations. What is the standard of time implied by the quantum mechanical equation of time evolution, and how is it related to the operational standard? Does the latter simply reproduce the undefined standard implied by the operator ∂_t, or is it in some way independent?

Before going on to resolve that issue, we should recall the argument by which the fundamental equation is usually established. It is that the wavefunction can and does incorporate all the possible information about the state of a system, including its evolution in time. It follows that the evolution of the wavefunction in time can only depend upon its value at some initial time, $t = 0$, and not on additional information such as the derivative, for that would imply that it was necessary to know the value at some time other than $t = 0$. If only one initial value is allowed, the equation for the evolution in time can only be of first order.

The postulate that the wavefunction contains all information about a system, is of course a formal element of the theory, and in practice some 'existence theorem' must be satisfied empirically, namely that systems can be found for which it is so. Quantum mechanics would not be so effective in giving an account of the world of empirical physics if there were not many systems to which the postulate does seem to apply.

2.2 An operational analysis

I now go back to look at the experimental operations of setting up the standard of frequency, to see how far they can be described in a way independent of conventional theory, what can be derived from them and what mathematical objects are needed to represent them.

The experiment is very simple in principle. Two states of the Cs-133 atom are defined by the magnetic selector, one, 'up', in which atoms are deflected in one direction and the other, 'down', in which they are deflected in the opposite direction. An electromagnetic field of appropriate frequency can change the state of the atom from the one to the other. That is all. How may it lead to an element of the mathematical structure of quantum mechanics?

In the first place, a quantum mechanical equation of time evolution has no meaning unless it refers to mathematical objects corresponding to physical states or processes, so it seems that we must postulate the existence of such objects that correspond to the two states of the Cs-133 atom in the apparatus. We wish to see if the experiment tells us how those mathematical objects would evolve in time.

Since we are dealing with time evolution in quantum mechanics, we naturally adopt the basic notions of the formalism of quantum mechanics, namely Hilbert spaces of wavefunctions that correspond to physical states, mathematical operators that correspond to changes of physical state and eigenvalues of operators that correspond to measurement on eigenstates. Our aim is to see how far the definition of our standard of time entails an equation of evolution in time for those mathematical objects postulated in the structure of quantum mechanics. We do not attempt to establish the structure itself which depends both on *a priori* arguments and historically upon the interaction between theory and experiment. A great deal is being taken as axiomatic and the aim is restricted.

The first step is to define the time scale of the experiment. It can only be the reciprocal of the transition frequency, for as yet in our restatement of the problem, no other time scale has been introduced. The particular experiment was chosen to realise the standard of frequency because it can do so with better precision – 1 part in 10^{13} – than any other scheme. The electromagnetic field thus has, as closely as can be realised, a single Fourier component, which means that the scale of time, the reciprocal of the frequency of that single component, is uniform and does not change with epoch, that is with elapsed time. Further, since there is no better standard, it is impossible to test experimentally whether the caesium standard of frequency varies with epoch. Indeed that is a meaningless question, the frequency can only be taken to be constant, and the time scale defined by the reciprocal of the standard frequency is a uniform scale.

It is trivial to observe that if V is some (complex) variable of the electromagnetic field, and if V_0 is its amplitude, so that

$$V = V_0 \exp(-\mathrm{i}\omega t)$$

then

$$\mathrm{i}\partial_t V = \omega V,$$

which is of the same form as

$$\mathrm{i}\hbar\partial_t \Psi = E\Psi.$$

We now see how far the first equation entails the second.

We first introduce a mathematical operator, D, corresponding to the second analyser and the detector in the experimental apparatus.

Let us give the label 1 to the state of atoms that pass the first analyser and the label 2 to that of atoms that pass the second analyser. When D operates on the wavefunction of state 1 it gives zero, while when it operates on the wavefunction of state 2 it multiplies that wavefunction by ρ, where ρ^2 is the ratio of the number of atoms falling on the detector to those in state 1:

$$D\Psi_1 = 0, \quad D\Psi_2 = \rho\Psi_2.$$

We also treat V as an operator, the effect of which is to transfer some atoms from the one state to the other over some brief interval of time, during which it may be arranged, and is commonly so arranged in the experiment, that the rates of change of the numbers of atoms in the two states are constants in time. Hence if we start with all atoms in state 1, some will be transferred into state 2 in that interval of time, while others will remain in state 1. We may therefore write the corresponding mathematical expression as

$$V\Psi_1 = a\Psi_2.$$

The factor a is a constant that includes terms corresponding to the geometrical arrangement of the experiment.

In general many atoms will not change their states and in the absence of the radiation V none will. We therefore need to postulate an identity operator I that leaves atoms in their original state. However, since the operator D annuls the wavefunction of state 1, it follows that the operator DI also gives a null result on operating on the wavefunction of state 1.

Overall, then, the experiment is described by the equation

$$DV\Psi_1 = a\rho\Psi_2$$

or

$$U\Psi_1 = \rho\Psi_2,$$

where U, equal to DV/a, represents the overall action of the apparatus.

The only factor of U that depends on time is the operator V and so U also satisfies the equation,

$$\mathrm{i}\partial_t U = \omega U.$$

Let us put $\Psi_1 = f_1(t)\psi_1$ and $\Psi_2 = f_2(t)\psi_2$, where f_1 and f_2 are functions of time only and ψ_1 and ψ_2 are functions of the coordinates but not of the time.

Thus

$$U f_1(t)\,\psi_1 = \rho f_2(t)\,\psi_2.$$

Take logarithms to give

$$\ln U + \ln f_1 + \ln \psi_1 = \ln f_2 + \ln(\rho\psi_2).$$

Now differentiate with respect to time. The functions ψ_1 and ψ_2 are independent of time so that $\partial_t \psi_1$ and $\partial_t \psi_2$ are zero, while $\partial_t \ln U = \partial_t U/U = -\mathrm{i}\omega$. Thus

$$-\mathrm{i}\omega + \partial_t f_1/f_1 = \partial_t f_2/f_2.$$

We therefore find that

$$\partial_t f_1 = A_1 f_1,$$
$$\partial_t f_2 = A_2 f_2.$$

where A_1 is an arbitrary function of space and time and A_2 is $(A_1 - \mathrm{i}\omega)$.

Thus for Ψ_1 and for Ψ_2

$$\partial_t \Psi = A\Psi.$$

The argument shows that the experimental method for the realisation of the standard of frequency entails a first order equation of time evolution for the wavefunctions representing each of the two states of the caesium atom. The operator denoted by A is obviously the analogue of the Hamiltonian in the usual form of the time-evolution equation, but we have not demonstrated that they are the same, only that there is some operator that gives the time rate of change. The form of that operator is not determined by any general argument but is essentially empirical, chosen to give agreement between calculation and observation. It is called the Hamiltonian because the choice of form is often guided by analogy with the Hamiltonian function of classical mechanics, but as we shall see in Chapter 4, classical mechanics is not always a safe guide. There are however circumstances in which a correspondence can be established between the classical function and the quantum operator, as will be seen below.

We have shown that two wavefunctions for the Cs-133 atom evolve in time according to the linear first order equation. What justification may there be for adopting the same equation more generally when experimental demonstration of equal force is not available for any but the standard system? We have shown that there is one process that follows the equation and therefore that in one case it can be said that there is a wavefunction that contains all the necessary information about the system. Whether or not other physical systems conform to the same model is a matter that has to be tested empirically, but two things can be said. In the first place, the frequencies of electromagnetic signals that cause transitions in other atomic or molecular systems can be related very precisely to the caesium standard and consequently the evolution of those systems in time can be placed on the common time scale of that standard.

Secondly, it can be asserted that so far the equation of time evolution, taken with the proper choice of Hamiltonian (an essentially empirical matter), has proved sufficiently general to handle almost all known physical phenomena. In particular, whenever electromagnetic fields of well-defined frequencies cause transitions between otherwise stable states, we can apply the above argument to show that the wavefunctions must satisfy the same equation of time evolution, with the scale of time identical to that of the caesium frequency standard.

Two arguments have been put forward for the first order equation of time evolution, the one based on the postulate that wavefunctions must contain all information necessary for the description of a physical system, the other based on the operational realisation of a standard of frequency. Neither can be shown to apply universally, for there may exist physical phenomena that cannot be fitted into the formalism of quantum mechanics; equally it cannot be shown that all phenomena necessarily entail the first order equation. Even though we know of no exceptions at present, the question of whether or not the formal requirements or the operational implications apply to any particular phenomenon is an empirical question, to be settled by comparing the prediction of quantum mechanics with the outcome of observation.

We have shown that as a result of the way in which we define the standard of frequency, certain wavefunctions have to satisfy a first order equation of time evolution. Formal arguments then tell us that all available information about the physical system is provided by those wavefunctions.

2.3 Classical standards of time and equations of motion

Standards of time realised through systems obeying classical dynamics are now obsolete, but it is still instructive to consider how they are related to the classical equations of motion.

In classical standards the fundamental interval of time, T, is that taken by some angular variable to increase by 2π. Thus ordinary sidereal time is defined by the successive instants at which the direction of the meridian of Greenwich points to some chosen fixed star, while ephemeris time is defined by the successive instants at which the Sun is seen from the Earth in the same direction relative to the fixed stars. In the first case the standard of time is realised through the rotation of the Earth upon its axis, while in the second it is realised through the motion of the Earth in its orbit about the Sun. As with the quantum standard, once the physical realisation has been chosen, the time interval defined by it is invariant by definition and the repetitions of the phenomenon establish a uniform scale of time – it is meaningless to contemplate possible variations of the scale with epoch since there is no independent way of finding out if they have occurred.

It follows then that if t_n and t_m are the times of the nth and mth occurrences of the phenomenon, then whatever the values of n and m,

$$t_n - t_m = (n-m)T.$$

We do not know what happens at times between the recurrences of the phenomenon but have to assume that between them the time scale is also uniform, so that, if ϕ is the angle that increases by 2π in time T, we have to set $\mathrm{d}\phi/\mathrm{d}t$ equal to a constant (a, say).

A single variable is insufficient to describe the state of any dynamical system and in addition to a coordinate we need a second independent variable that depends on the mass and perhaps other properties of the system. Let us call it $\mathcal{J}$. We may then write formally

$$\mathrm{d}\phi/\mathrm{d}t = \mathrm{d}(a\mathcal{J})/\mathrm{d}\mathcal{J}.$$

But because $\mathrm{d}\phi/\mathrm{d}t$ is a constant, $\mathrm{d}^2\phi/\mathrm{d}t^2$ is zero and therefore so also is $\mathrm{d}(a\mathcal{J})/\mathrm{d}t$ and thus $\mathrm{d}\mathcal{J}/\mathrm{d}t$. Furthermore, $a\mathcal{J}$ does not depend on ϕ, so $\mathrm{d}(a\mathcal{J})/\mathrm{d}\phi$ is zero and thus

$$\mathrm{d}\mathcal{J}/\mathrm{d}t = \mathrm{d}(a\mathcal{J})/\mathrm{d}\phi.$$

If we set $-a\mathcal{J}$ equal to a function $K(\phi,\mathcal{J})$, of ϕ and $\mathcal{J}$ we have formally

$$\mathrm{d}\mathcal{J}/\mathrm{d}t = \partial K/\partial\phi, \quad \mathrm{d}\phi/\mathrm{d}t = -\partial K/\partial\mathcal{J},$$

with the solutions, $\phi = at$, $\mathcal{J} = \text{constant}$, because $\partial K/\partial\phi = \mathrm{d}(a\mathcal{J})/\mathrm{d}\phi$ which is zero and $\partial K/\partial\mathcal{J} = -a$.

We see therefore that the definition of time by a classical periodic system means that there is one system at least that evolves in time according to the Hamiltonian equations of motion. We emphasise again that the argument establishes the form of the equations of motion but not the content of the function K which is analogous to the Hamiltonian but not demonstrably identical.

If we are dealing with the rotation of the Earth upon its polar axis, then $\mathcal{J}$ is the spin angular momentum of the Earth about the polar axis, while if we consider the motion of the Earth in its orbit about the Sun, $\mathcal{J}$ is the orbital angular momentum of the Earth about the Sun. The motion of a pendulum can also be described by an angle which increases uniformly with time, and so can be used to define a scale of time, because the angular position of the pendulum is proportional to the cosine of a phase angle which is equal to the frequency of the pendulum multiplied by the time.

An obvious question is whether the classical and quantum scales of time agree; it is essentially an empirical issue. It is well known that it can be shown in various ways that quantum mechanics goes over in to classical mechanics when h is small relative to the action of a system. One way is to take appropriate averages over the quantum mechanical variables (see for example Fröhlich, 1967, 1973). Another route depends on the correspondence between the quantum mechanical wavefunction, Ψ and the classical action, S, namely $\Psi = \exp(-\mathrm{i}S/\hbar)$, according to which the Schrödinger equation reduces to the Hamilton–Jacobi equation in the limit of h small compared to S. A third method relates classical and quantum quantities though the Weyl transformation (Ozorio de Almeida, 1988). If K is a quantum mechanical operator, its Weyl transform, K, is

$$\mathrm{K}(\mathbf{p},\mathbf{q}) = \int \mathrm{d}\mathbf{y}\langle \mathbf{q}+\mathbf{y}/2|K|\mathbf{q}-\mathbf{y}/2\rangle \exp(-\mathrm{i}\hbar\mathbf{p}\mathbf{y}).$$

Thus if K is the operator $\mathbf{q}$, $\mathrm{K}(\mathbf{p},\mathbf{q})$ is the coordinate $\mathbf{q}$, while if K is the operator $\mathbf{p}$, $\mathrm{K}(\mathbf{p},\mathbf{q})$ is the momentum $\mathbf{p}$.

If the general quantum mechanical equations for two bodies in mutual orbit are set up, it may be shown by any of the three methods that the classical equations of orbital motion follow from them. The time variable is left unaltered so that the time defined by the classical motion should bear an invariant relation to the quantum standard. That ignores the effects of pertubations by other bodies in the classical system and the possibility that they would cause deviations in the time scales if full allowance is not made for them. We know also that ephemeris time has to be corrected for the metric of general relativity (see the Appendix), and should there be other deviations of the gravitational attraction from the simple Newtonian form, then there would be further discrepancies between the quantum and mechanical scales.

2.4 Defining constants of physics

In the present international system of units many standards other than those of time or frequency are related to frequency by defining physical relations; they are no longer realised by independent physical means (Petley, 1985). Thus the standard of length is no longer realised by a bar of platinum–iridium alloy nor by the wavelength of light in the spectrum of krypton-86, but is defined in terms of the time taken by an electromagnetic signal to traverse some distance. The physical standard of length has been replaced by a specified speed of light which is necessarily, by definition, a constant. Constants such as the speed of light are often spoken of as fundamental constants, but it seems better to call them defining constants because through them the standard quantities of length, energy and electromagnetism are related to frequency. To speak of defining constants avoids any implication that the particular constants are logically fundamental in the formal structure of physics, whereas they were selected for reasons of metrological precision and convenience. The speed of light was chosen to define length because that is the most practical and precise way of making measurements of length over any distance greater than a few metres. Similarly, as we shall now see, the definitions of electrical quantities are effected most precisely and conveniently by practical relations to frequency. Apart from precision, an important consideration is that standards of frequency are widely and accurately disseminated by radio broadcasts, so that once the relation of frequency to some other quantity is accepted, the standard of that other quantity is just as accessible as the standard of frequency (although in general the other standards cannot be realised as precisely as the standard of frequency).

For some while now the standard of electrical potential has been related to frequency through the Josephson effect that occurs between two superconducting elements joined by a weakly conducting link. Let a dc potential difference be maintained across the link, and let the junction be placed in a radio frequency electromagnetic field. The relation of the dc current through the link to the applied voltage is then a series of steps and the magnitudes of the increments of voltage from step to step have been shown by careful experiments to be independent of the materials of the junction within experimental error and to be given by

$$h\nu = 2eV,$$

where ν is the frequency of the rf field, h is Planck's constant and e is the charge on the electron. Irrespective of the physical significance of h and e, the ratio $h/2e$ is a defining constant in the sense introduced above.

The relation states in effect that the energy absorbed from the field is equal to that required for a pair of electrons to pass from one side of the junction to the other against the potential difference.

The quantum Hall effect gives a definite relation between current and voltage in particular circumstances. It occurs at temperatures below 4K in two-dimensional solid state structures in which electrons are constrained to move in plane orbits when a high magnetic field is applied perpendicular to the plane of the structure. If a current is driven through the structure in one direction in the plane, x say, then a potential difference is developed in the perpendicular direction in the plane (the y-direction), just as in the normal Hall effect. However, at certain magnetic fields the direct resistance, R_{xx}, in the x-direction falls to zero, but the Hall resistance, R_{xy}, the ratio of voltage across the y-direction to current in the x-direction, takes on successively values equal to h/ne^2, where n is a small integer. The value of h/e^2, which is denoted by R_{H}, is about 25813 ohm.

Evidently if voltage is related to frequency through the

Josephson effect as $V = h\nu/2e$, then a current can be determined from that voltage and related to frequency through the formulae

$$i = V/R_H = Ve^2/nh = 2e\nu/n.$$

Further, power, which is Vi, or V^2/R_H, is $h\nu^2/4n$.

In the local laboratory a set of observations in which a secondary standard of voltage is derived from the Josephson effect and a secondary standard of resistance from the quantum Hall effect, enables any voltage, current or resistance to be related to frequency through the respective constants $h/2e$, h/eR_H, equal to e, and h^2/e^2R_H equal to h. Notice that h/eR_H is equal to e as it obviously must be, since $e\nu$, the rate of transport of charge, is current. Similarly h^2/e^2R_H reduces to h since power is rate of change of rate of change of action, namely $h\nu^2$. However, the constants that are most accurately determined in terms of the older electrical standards of voltage and current are not h and e but the combinations $h/2e$ and R_H.

We know of no way of relating the standard of mass directly to frequency and at present it is still defined by the mass of a particular cylinder of platinum–iridium alloy conserved in the Bureau International des Poids et Measures at Sèvres. An alternative now presents itself however, to make use of the relation of power to frequency that can be effected through the Josephson effect and the quantum Hall effect. If a certain mass, m, is given a certain velocity v in time t, it will have acquired kinetic energy $mv^2/2$ which will be equal to the integral of the power expended in accelerating it over the time t. So far no means of doing that with adequate precision has been devised, but in principle it is possible to abandon the metal cylinder as a standard of mass.

Alternatively, it would be possible to take force instead of mass as a fundamental quantity, defining it perhaps in the reverse way to that by which the Ampère was defined through the force between two conductors at a specified separation carrying given currents.

2.5 Summary

The standard of frequency is defined by an operation in atomic physics whereby atoms are transferred from one well-defined state to another when subject to electromagnetic radiation at a very precisely determined frequency, the standard frequency. We suppose that in accordance with the basic postulate of quantum mechanics the two states correspond to two wavefunctions. The nature of the experimental operation then implies that each wavefunction changes with time according to the first order time-evolution equation,

$$\partial_t \Psi = H\Psi.$$

If we may adopt the basic postulate that a physical state corresponds to a wavefunction, then the wavefunctions corresponding to the two hyperfine states of the caesium-133 atom evolve with time according to that equation. We further see that because the solution of the equation depends upon an initial value of the wavefunction at one time only, the wavefunction determines the whole evolution of the atomic system from that time onwards.

Once a standard frequency has been defined, operations of quantum physics enable us to relate most other physical quantities to frequency through a few well-established defining constants. In particular, distance is now defined in terms of frequency and the speed of light, and that enables us to make measurements of time and distance to and between events far removed from us. The geometrical consequences of making measurements in that way are derived in the next chapter.

3

Observations at a distance: special relativity

3.1 Introduction

VARIOUS WAYS OF establishing the principles and transformations of special relativity have been appealed to in the past. As a matter of history, the Lorentz transformation was constructed to ensure that Maxwell's equations for the electromagnetic field did not change in form when the coordinates of time and distance were changed. If we put all arbitrary constants equal to unity, Maxwell's equations for the **E** and **B** fields in free space have the form

$$
\begin{aligned}
\mathrm{div}\mathbf{E} &= 0, \\
\mathrm{div}\mathbf{B} &= 0, \\
\mathbf{curlE} &= -\partial_t \mathbf{B}, \\
\mathbf{curlB} &= \partial_t \mathbf{E}.
\end{aligned}
$$

Time as well as space derivatives enter the second pair of equations, so that if the operators ∂_t and ∂_x are related by the velocity of the coordinate frame relative to the field, a transformation of the space coordinates will change the time derivative. If the form of the equations is not to be altered by a shift of coordinates, as Maxwell himself saw was necessary, the time derivative must change correspondingly. The Lorentz transformation is the transformation that maintains the form of Maxwell's equations.

The result is seen most clearly by using the 4-potential, the components of which, in tensor notation, are the scalar potential A_0 and the three components of the vector potential, A_i ($i = 1$–3). In the scheme of special relativity, the 4-potential is a 4-vector and transforms according to the Lorentz transformation.

Let $\mathbf{v}$ be the velocity of an observer relative to the field, and choose the spatial coordinates so that the 1(or i)-direction is parallel to $\mathbf{v}$. Taking the time coordinate to be x^0 the matrix of the transformation is

$$\begin{pmatrix} \cosh\zeta & \sinh\zeta & . & . \\ \sinh\zeta & \cosh\zeta & . & . \\ . & . & 1 & . \\ . & . & . & 1 \end{pmatrix} \text{ or alternatively } \begin{pmatrix} \gamma & \gamma v & . & . \\ \gamma v & \gamma & . & . \\ . & . & 1 & . \\ . & . & . & 1 \end{pmatrix}$$

where ζ is arctanh (v/c) and γ is $(1-v^2/c^2)^{-1}$; thus $\cosh\zeta$ is γ and $\sinh\zeta$ is γv. (v^2 is the square of the magnitude of the vector $\mathbf{v}$).

The transformed potentials are accordingly

$$\begin{aligned} A'_0 &= A_0\cosh\zeta + A_i\sinh\zeta \\ A'_i &= A_i\cosh\zeta + A_0\sinh\zeta \\ A'_j &= A_j, \quad A'_k = A_k; \end{aligned}$$

or

$$\begin{aligned} A'_0 &= \gamma(A_0 + vA_i), \\ A'_i &= \gamma(A_i + vA_0), \\ A'_j &= A_j, \quad A'_k = A_k. \end{aligned}$$

In a similar way the differential operators transform as

$$\begin{aligned} \partial'_0 &= \partial_0\cosh\zeta + \partial_i\sinh\zeta \\ \partial'_i &= \partial_i\cosh\zeta + \partial_0\sinh\zeta \\ \partial'_j &= \partial_j, \quad \partial'_k = \partial_k; \end{aligned}$$

or

$$\begin{aligned} \partial'_0 &= \gamma(\partial_0 + v\partial_i), \\ \partial'_i &= \gamma(\partial_i + v\partial_0), \\ \partial'_j &= \partial_j, \quad \partial'_k = \partial_k. \end{aligned}$$

The magnetic field is **curlA**, with components B_i equal to $(\partial_j A_k - \partial_k A_j)$, and so on, while those of the electric field, E_i, are $(\partial_i A_0 - \partial_0 A_i)$.

The transformed components are obtained by replacing the unprimed by primed potentials and operators, so that $B'_i = B_i$, while

$$\begin{aligned} B'_j &= (\partial_i A_k - \partial_k A_i)' \\ &= (\partial_i \cosh\zeta + \partial_0 \sinh\zeta)\, A_k - \partial_k(A_i \cosh\zeta + A_0 \sinh\zeta) \\ &= B_j \cosh\zeta + E_k \sinh\zeta, \end{aligned}$$

Similarly

$$B'_k = B_k \cosh\zeta - E_j \sinh\zeta.$$

Alternatively

$$B'_i = B_i; \quad B'_{j,k} = \gamma[\mathbf{B} + (\mathbf{E} \wedge \mathbf{v})]_{j,k}.$$

The expressions for the j and k components of the transformed electric field follow in a similar way and have the same form as for the magnetic field components.

The ith component of the transformed electric field is

$$E'_i = \partial'_i A'_0 - \partial'_0 A'_i,$$

that is

$$(\partial_i \cosh\zeta + \partial_0 \sinh\zeta)\,(A_0 \cosh\zeta + A_i \sinh\zeta) \\ - (\partial_0 \cosh\zeta + \partial_i \sinh\zeta)\,(A_i \cosh\zeta + A_0 \sinh\zeta),$$

which is E_i because $\cosh^2\zeta - \sinh^2\zeta = 1$.

We can now go on to transform Maxwell's equations. Consider first $\{(\mathbf{curl}'\mathbf{B}')_i - \partial'_0 E'_i\}$, the ith component of the left side of the transformed equation

$$\mathbf{curl}\mathbf{B} - \partial_0 \mathbf{E} = 0.$$

The component is $\partial'_j B'_k - \partial'_k B'_j - \partial'_0 E'_i$ and it transforms to

$$\gamma\partial_j(B_k - vE_j) - \gamma\partial_k(B_j + vE_k) - \gamma(\partial_0 + v\partial_i)\, E_i,$$

that is

$$\gamma(\mathbf{curl}\mathbf{B} - \partial_0 \mathbf{E})_i - \gamma v\, \mathrm{div}\mathbf{E},$$

which vanishes because $(\mathbf{curl}\mathbf{B} - \partial_0 \mathbf{E})$ is identically zero and $\mathrm{div}\mathbf{E}$ is zero in free space.

We also have

$$\mathrm{div}'\mathbf{B}' = \partial'_i B'_i + \partial'_j B'_j + \partial'_k B'_k,$$

that is

$$\gamma(\partial_i + v\partial_0)\, B_i + \gamma\partial_j(B_j + vE_k) + \gamma\partial_k(B_k - vE_j)$$

or

$$\gamma\, \mathrm{div}\mathbf{B} + \gamma v\{\partial_0 \mathbf{B} + \mathbf{curl}\mathbf{E}\}_i,$$

which is zero because both div**B** and $\{\partial_0 \mathbf{B} + \mathbf{curl E}\}_i$ vanish identically.

Likewise we may show that div**E** and $(\mathbf{curl E} + \partial_0 \mathbf{B})$ remain zero under the Lorentz transformation.

Let us write the square of the space-time interval as

$$s^2 = (x^0)^2 - r^2$$

where $r^2 = (x^1)^2 + (x^2)^2 + (x^3)^2$. s^2 is invariant under the Lorentz transformation. Einstein based his development of special relativity on that result and extended it to the invariance of the energy-momentum 4-vector with components $(\mathbf{E}, c\mathbf{p})$, where $\mathbf{E}$ is the energy and $\mathbf{p}$ the 3-momentum. Thus for a single particle

$$E^2 - c^2 p^2 = E_0^2,$$

where E_0 is the invariant rest energy of a particle at rest, namely $m_0 c^2$.

One way in which the invariance of the interval s is established is to assume that the speed of light is constant and to observe that on a light path, the space-like interval $\mathrm{d}r$, when measured as the time of travel of a light signal, is identical with the time interval $\mathrm{d}x^0$ (here as elsewhere, all distances are expressed as transit times of light signals, and the speed of light, taken to be unity, is suppressed). The scalar difference $(\mathrm{d}x^0 - \mathrm{d}r)$ vanishes for two events connected by a light path. That will remain true whatever transformations of the coordinates are applied. Arguments based on the homogeneity and consistency of transformations (see Landau and Lifschitz, 1971) show that if

$$\mathrm{d}s^2 = (\mathrm{d}x^0)^2 - \mathrm{d}r^2$$

remains zero under some transformation, then so does the interval between any other two events which are not connected by a light path, provided that the ratio $(\mathrm{d}r/\mathrm{d}x^0)$ remains constant in the transformation, that is, the speed of light is unaltered.

The usual way of establishing the rules of special relativity thus depends on the assumption that the speed of light is constant and on arguments about the consistency of transformations. We shall now go on to show that the rules of special relativity follow from the operational procedures of metrology, just as the quantum mechanical equation of time evolution was seen to be the consequence of the way in which the standard of frequency is defined.

3.2 Geometrical observations at a distance

All the following arguments are founded on the fundamental fact that the only way of making observations at a distance across free space is with electromagnetic waves. One can observe the times of emission and reception of electromagnetic signals at her own position, the angles between the directions in which signals are launched or received, and the frequencies and polarisations of signals. No other observations of distant events are possible. We must first discuss what we mean by the constancy of the speed of light in those circumstances and we must also consider what meaning we can give to the space-time coordinates of distant events that we can observe only by means of electromagnetic signals.

We note first that the scale of time in which all our measurements will be expressed is that established by the observer's standard of time (or frequency) in her own laboratory. We might consider making a measurement of the frequency of a similar standard of frequency operating at a distance, but we cannot in reality do so, we can only observe the frequency of the distant standard as it is measured by us having been transmitted to us by an electromagnetic signal.

There is a second observation that we cannot make, the determination of the speed of light, for we have no standards of length, independent of transit times of light waves, that we can use over great distances. Indeed the standard of length has been defined in terms of the time of passage of light signals because, among other reasons, independent standards of length cannot be set up for any distance greater than a few metres.

Since it is impossible to measure directly the frequency of a distant standard oscillator, we can only treat it as equal to that of a similar standard oscillator in our own local laboratory; the frequencies of similar standard oscillators must be supposed to be the same wherever they are situated. A corollary of that is that all times of epoch and of transit have to be expressed in terms of the observer's local time scale. We must also assume that the speed of light is constant, for if we cannot establish an independent standard of length, we cannot test whether the speed of light is or is not constant. Those two assumptions, the invariant frequency of the caesium standard and the constancy of the speed of light, may perhaps seem arbitrary, but we have to adopt them because there is nothing else we can do, we have no way of checking them, at least locally, and by adopting them we recognise that inability. Arbitrary assumptions correspond to impossible observations. In fact it turns out (Section 3.5) that there are astronomical observations that are inconsistent with the assumption that the speed of light is constant, and special relativity has to be extended to general relativity to deal with that. For the present we restrict ourselves to a constant speed of light.

Let two events be specified each by a time and a position, in some system of time and space coordinates. Let $\mathrm{d}r$ be the time of transit of an electromagnetic signal between the positions of the two events, $\mathrm{d}r$ is the modulus of the vector separation of the events, $\mathrm{d}\mathbf{r}$. We have already observed that if the time interval, $\mathrm{d}x^0$ between the events is equal to $\mathrm{d}r$, the events are joined by a light path; the quantity $\mathrm{d}s = (\mathrm{d}x^0 - \mathrm{d}r)$ is zero for events joined by a light path. We now show that when any two events are observed from a distance, then the only quantity that can be derived from the possible observations is $\mathrm{d}s$, or rather, $\mathrm{d}s^2$, measured in units of the observer's standard of time.

Consider first, how to combine times of transit. In the immediate locality of an observer they are proportional to Euclidean distances and so combine according to Euclidean metrology; that is the basis of surveying performed by measuring the separations of intervisible stations through electromagnetic signalling between them. When we come to deal with places at great distances, it might be argued that since we have no direct knowledge of the appropriate geometry, we can only assume it to be Euclidean. In fact, we can make a stronger argument. Euclidean geometry deals with abstract quantities that we call lines and points and for which we have axioms of incidence. Objects in familiar three-dimensional space correspond to the lines and points of Euclidean geometry. We also need the definition of the length of a line, that is of the norm of a vector, and also the definition of the scalar product of two vectors. From those definitions there follows, for example, Pythagoras's theorem. Ray paths of electromagnetic radiation and receivers and transmitters of signals (idealised as point objects) also satisfy the Euclidean axioms of incidence. It remains to define a metric for the geometry and here the key point is that if the speed of light is supposed to be constant throughout space, then the metric coefficients are just constants and so the square of the interval between two points is given by Pythagoras's theorem. That is equivalent to saying that if the speed of light is a constant independent of position, then the paths of light rays in free space are straight lines without curvature.

We have already seen that the speed of light has to be taken as a universal constant, and that it has been incorporated in the international definition of the standard of length, and we now see that there is more to that postulate than just the value of the conversion factor from light times to metres, we see that upon it depends the use of the geometry of straight lines for combining travel times. General relativity, discussed further in Section 3.5, is the generalisation of the geometry in which the condition that ray paths are straight lines is relaxed and the curvature is supposed to depend on the potential of matter in the universe.

It will now be clear how to calculate the transit time of an electromagnetic signal between two distant points from the times of travel of signals from an observer to each point – it is to be found from the cosine rule for plane triangles.

The angle between the rays joining an observer to two events is found from measurements of differences of times of travel. In order to obtain the direction of a ray to a distant event in the observer's frame of reference, the event has to be observed from two points that define a base line of known length and orientation in the observer's frame (Figure 3.1). Let the ends of such a base line (of length $2D$, say) be labelled a and b. Let the transit time of a signal from the mid-point of the base line to an event labelled 1 be T_1 and let the times from the ends of the base line be T_{1a} and T_{1b}. Let ϕ_1 be the angle between the normal to the base line and the direction of event 1 from the mid-point of the base line. Then by the cosine rule, taking ϕ_1 to be negative,

$$T_{1a}^2 = T_1^2 + D^2 + 2DT_1 \sin\phi_1$$

and

$$T_{1b}^2 = T_1^2 + D^2 - 2DT_1 \sin\phi_1.$$

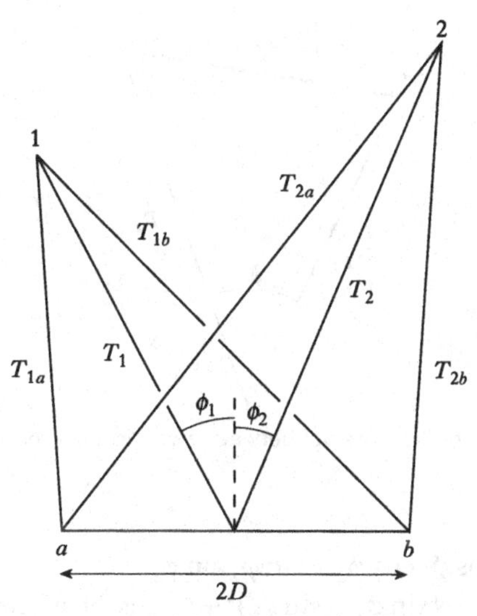

Figure 3.1. Observation of direction by electromagnetic signals.

Now the times T will in practice be very much greater than the time D and consequently

$$\sin \phi_1 = (T_{1a} - T_{1b})/2D,$$

while

$$T_1 = (T_{1a} + T_{1b})/2.$$

On the face of it, that description of how the direction of a source is found from differences of travel times applies to radio observations in which time differences are measured directly. It is not in fact so restricted. When the direction of a source is determined by interferometric measurements at whatever frequency, the observed phase differences correspond to time differences, and even the observation of direction by pointing with an ordinary telescope depends on phase difference, and therefore time difference, across the aperture of the telescope.

The transit time dr for a signal going between events 1 and 2 follows from the cosine rule:

$$\mathrm{d}r^2 = T_1^2 + T_2^2 - 2T_1 T_2 \cos \phi.$$

where ϕ is $(\phi_2 - \phi_1)$.

Hence

$$\mathrm{d}r^2 = (T_2 - T_1)^2 + 2T_1 T_2(1 - \cos \phi).$$

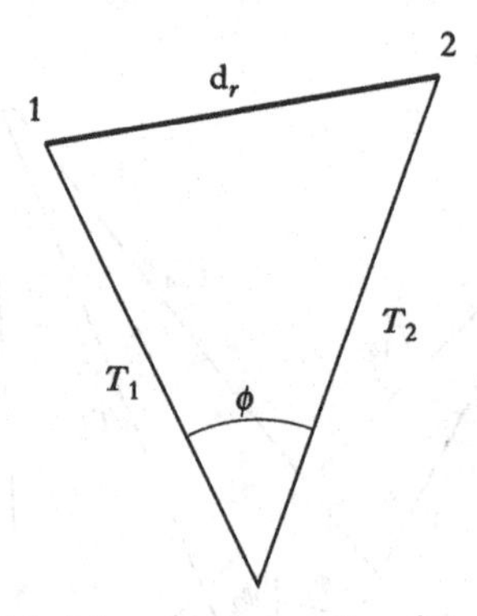

Figure 3.2. Interval between two distant events.

But

$$\cos\phi = \cos\phi_1\cos\phi_2 + \sin\phi_1\sin\phi_2$$
$$= 1 - \tfrac{1}{2}(\sin\phi_1 - \sin\phi_2)^2 + \text{terms of higher order},$$

and so

$$1 - \cos\phi = (\mathrm{d}T_1 - \mathrm{d}T_2)^2/8D^2,$$

where

$$\mathrm{d}T_1 = (T_{1a} - T_{1b}), \quad \mathrm{d}T_2 = (T_{2a} - T_{2b}).$$

Thus finally,

$$\mathrm{d}r^2 = (T_2 - T_1)^2 + T_1T_2(\mathrm{d}T_1 - \mathrm{d}T_2)^2/4D^2.$$

Suppose first that two events are in line as seen by an observer. The transit time dr from 1 to 2 as calculated by the observer is then just $(T_2 - T_1)$.

Suppose that in order for a signal to reach event 2 it has to be launched by the observer at a time t after the launch of the signal that reaches event 1. If 1 and 2 are on a light path, t is zero, for the signal that reaches 1 will continue on to reach 2, but if event 2 occurs after the signal from the observer to 1 would reach the position of 2, t is greater than zero. On a light path

$$\mathrm{d}x^0 = T_2 - T_1$$

but generally

$$\mathrm{d}x^0 = T_2 - T_1 + t.$$

If the events 1 and 2 are not in line as seen from the observer (Figure 3.2) the transit time between them has to be found from the cosine rule:

$$\mathrm{d}r^2 = T_1^2 + T_2^2 - 2T_1T_2\cos\phi,$$

and thus it is the square of a transit time that has to be combined with a time interval. On a light path we have $(\mathrm{d}x^0)^2 = \mathrm{d}r^2$, and $\mathrm{d}s^2$ is zero, while for events not on a light path we consider $\mathrm{d}s^2 = (\mathrm{d}x^0)^2 - \mathrm{d}r^2$.

When the events are in line with the observer,

$$\begin{aligned} \mathrm{d}s^2 &= (T_2 - T_1 + t)^2 - (T_2 - T_1)^2 \\ &= t\{2(T_2 - T_1) + t\}. \end{aligned}$$

That is an expression that depends only on the observations and is independent of any assumption about or knowledge of the position or velocity of the distant events. Consequently when coordinates are assigned to the distant events, it must be done in such a way that the value of $\mathrm{d}s^2$ is unchanged. This is the crucial argument. We have found a mathematical object, namely

$$t\{2(T_2 - T_1) + t\},$$

that is a function only of the observations and that reduces to zero for events joined by a light path, that is when t is zero. Because it depends only on the observations it cannot be allowed to alter and any system of coordinates must leave it invariant.

When the events are not in line from the observer there is an obvious modification to the transit time. By the cosine rule:

$$\mathrm{d}r^2 = (T_2 - T_1)^2 + 2T_1 T_2(1 - \cos\phi);$$

$\mathrm{d}x^0$ however is unaltered.

Thus

$$\begin{aligned} \mathrm{d}s^2 &= (T_2 - T_1 + t)^2 - (T_2 - T_1)^2 - 2T_1 T_2(1 - \cos\phi) \\ &= t\{2(T_2 - T_1) + t\} - T_1 T_2(\mathrm{d}T_1 - \mathrm{d}T_2)^2/4D^2, \end{aligned}$$

again an expression that is a function of the observations alone.

It will be clear from that account of the measurement of time differences, whether by radio means or optically, that the results depend heavily on technique, especially electronic methods for detection of weak signals and on very precise time and frequency standards, and that they also depend heavily on theory, especially on the theory of the interference of electromagnetic waves and the operation of interferometers.

It remains to discuss the meaning of coordinates of distant events and how they are to be assigned. The elementary situation which confronts us is that we make observations of two events in a distant frame that is at some unknown distance from us and moving at some unknown velocity relative to us. We can derive the square of the space-time interval between the events from the observations – that is the physical reality. Coordinates are a frame imposed upon the events that we have no way of observing directly; they are arbitrary, but different sets of coordinates must be related in such a way that the observational result, the square of the space-time interval, is preserved. Coordinates have in general no physical reality nor significance but are assigned in the distant field for convenience in thinking about and doing calculations on the geometry of that field. Not only must they preserve the value of the space-time interval ds but they must preserve scale between space-like and time-like intervals, that is to say, the speed of light must be kept unchanged.

That view of coordinates in a distant field differs from the discussion of Toraldo di Francia (1981) who considered in effect that there is no real distinction between observational and theoretical quantities and that all can be defined operationally. The distinction here is that the possibility of variation is incorporated in the definition of coordinates (and similar quantities such as potentials and fields) from the start. Toraldo di Francia's argument could perhaps be accepted if it were taken to apply to the whole class of coordinates equivalent under a Lorentz transformation and not to one particular scheme.

Suppose that the distant field is moving relative to the observer with velocity $\mathbf{v}$. Take Cartesian coordinates in the distant field such that the x^i-axis coincides with the direction of $\mathbf{v}$. The matrix of the transformation that satisfies the conditions is, as in Section 3.1,

$$\begin{pmatrix} \cosh\zeta & \sinh\zeta & . & . \\ \sinh\zeta & \cosh\zeta & . & . \\ . & . & 1 & . \\ . & . & . & 1 \end{pmatrix},$$

where as before, $\tanh\zeta$ is v/c.

Readers may easily satisfy themselves that if the original coordinate intervals are $(\mathrm{d}x^0, \mathrm{d}x^1, \mathrm{d}x^2, \mathrm{d}x^3)$ with a space-time interval ds, then the space-time interval ds' of the transformed set $(\mathrm{d}x^{0\prime}, \mathrm{d}x^{1\prime}, \mathrm{d}x^{2\prime}, \mathrm{d}x^{3\prime})$ is equal to the original ds.

Sets of four quantities that transform by the Lorentz transformation and maintain the same magnitude are 4-vectors. In the particular case of space-time geometry, the magnitude is the space-time interval as already defined, and the mathematics has to be devised to ensure that it is constant because it is the result of observation. If some other set of four quantities corresponds to a set of physical observations, its magnitude, considered as that of a 4-vector, must also be a constant and equal to an appropriate combination of the observations and of them alone. The property of 4-vectors, that their scalar products are invariant under Lorentz transformations, enables the magnitudes of 4-vectors to be identified with the results of physical measurements, which must remain unchanged in any theory. It will be recalled that for the difference of the squares of the components of a vector to be invariant, the metric of special relativity, with which the scalar product is formed, has to be the diagonal matrix $\{1,-1,-1,-1\}$.

We now go on to consider 4-vectors that represent other than geometrical observations.

3.3 4-vectors as representing physical observations

Let us begin with a 4-vector that is related to the metric properties of electromagnetic radiation. Radiation that has a single Fourier component has a unique phase which can be measured by an observer, for when the signal falls to zero, that condition can be detected by an observer and is independent of the velocity of the observer, it is a property of the signal alone. Consequently the value of the phase as calculated in any theory must not depend on the position or velocity of the observer.

Let the observer's coordinates be $(x^0, \mathbf{r})$ where $\mathbf{r}$ is as before the 3-vector of spatial position. Let $\mathbf{k}$ be the wave vector of the signal (a 3-vector). Then the phase is $(\omega x^0 - \mathbf{k}\cdot\mathbf{r})$, where ω is the circular frequency of the signal.

Frequency and wavevector are of course measured in the observer's frame of reference to correspond with the coordinates.

Since phase is an observed quantity it must be independent of coordinate transformations. It is a linear combination of frequency and wavevector with the coordinates, and it must be a scalar product of 4-vectors if it is to be independent of coordinate transformations. The coordinates $(x^0, \mathbf{r})$ constitute a 4-vector and so frequency and wavevector must also form a 4-vector. That vector is in fact a null vector, for the magnitude of the wavevector is equal to the frequency in units in which the speed of light is unity, that is, $\omega^2 - k^2 = 0$.

In the particular case in which $\mathbf{k}$ is in the direction of the velocity of the observer's frame, the transformation is

$$\begin{pmatrix} \omega' \\ k_1' \\ k_2' \\ k_3' \end{pmatrix} = \begin{pmatrix} \cosh\zeta & \sinh\zeta & . & . \\ \sinh\zeta & \cosh\zeta & . & . \\ . & . & 1 & . \\ . & . & . & 1 \end{pmatrix} \begin{pmatrix} \omega \\ k_1 \\ 0 \\ 0 \end{pmatrix},$$

so that

$$\omega' = \omega\cosh\zeta + k_1\sinh\zeta,$$
$$k_1' = \omega\sinh\zeta + k_1\cosh\zeta,$$
$$k_2' = k_3' = 0$$

($\tanh\zeta$ is as usual v/c).

The expressions are just those for the relativistic Doppler shift under the specified conditions. We again take the speed of light to be unity and again put

$$\cosh\zeta = \gamma = (1-v^2)^{\frac{1}{2}}, \quad \sinh\zeta = \gamma v.$$

Then

$$\omega' = \gamma\omega + \gamma v k_1, \quad k_1' = \gamma v\omega + \gamma k_1,$$

which are the formulae for the relativistic Doppler shift.

Now k_1 equals ω in units in which c is 1 and so

$$\omega' = \gamma\omega(1+v) = k_1'.$$

We see that the transformed frequency and wavevector are equal in magnitude, as is necessary since the speed of light is unchanged and frequency-wavevector 4-vector must remain a null vector.

In a second simple case the wavevector is perpendicular to the velocity of the observer's frame. Let the source be in the direction 2 so that the only non-zero component of the wavevector is k_2. Then the Lorentz transformation gives

$$\begin{pmatrix} \omega' \\ k_1' \\ k_2' \\ k_3' \end{pmatrix} = \begin{pmatrix} \cosh\zeta & \sinh\zeta & . & . \\ \sinh\zeta & \cosh\zeta & . & . \\ . & . & 1 & . \\ . & . & . & 1 \end{pmatrix} \begin{pmatrix} \omega \\ 0 \\ k_2 \\ 0 \end{pmatrix},$$

whence

$$\omega' = \omega \cosh \zeta, \quad k_1' = \omega \sinh \zeta$$
$$k_2' = k_2, \quad k_3' = 0.$$

Then $\omega' = \gamma\omega$, the relativistic Doppler shift, while k_1' is equal to $\gamma v \omega = \gamma v k_2$, and again ω', which is $\{(k_1')^2 + (k_2')^2\}^{\frac{1}{2}}$, is equal to the norm, k'.

The direction of the source as it appears to the observer has direction cosines proportional to k_1' and k_2', and since k_2' is equal to k_2, the apparent direction of the ray changes by the angle $\arcsin(\gamma v)$, the relativistic aberration.

If we differentiate the space-time vector d**s** with respect to its magnitude ds, we obtain an object with components $(\partial x^0/\partial s, \partial \mathbf{r}/\partial s)$. It obviously transforms as a 4-vector under the Lorentz transformation because the vector $(dx^0, d\mathbf{r})$ transforms in that way and the magnitude, ds, is invariant under the transformation.

The square of the magnitude of $\partial \mathbf{s}/\partial s$ is

$$\left(\frac{\partial x^0}{\partial s}\right)^2 - \frac{\partial \mathbf{r}}{\partial s}\cdot\frac{\partial \mathbf{r}}{\partial s}.$$

Now

$$\frac{\partial \mathbf{r}}{\partial s} = \frac{\partial \mathbf{r}}{\partial x^0}\frac{\partial x^0}{\partial s} = \mathbf{v}\frac{\partial x^0}{\partial s}.$$

Thus

$$\left(\frac{\partial \mathbf{s}}{\partial s}\right)^2 = \left(\frac{\partial x^0}{\partial s}\right)^2 (1 - v^2) = 1,$$

because $(\partial x^0/\partial s)^2$ is $\cosh^2 \zeta$ which is $(1 - v^2)^{-1}$.

The magnitude of the 4-velocity is therefore 1 when the speed of light is taken to be unity.

Since $(\partial x^0/\partial s)$ is the same as γ, the components of the 4-velocity are $\gamma(1, -\mathbf{v})$.

A 4-acceleration may also be defined as $\partial(\partial \mathbf{s}/\partial s)/\partial s$, that is

$$\left\{\frac{\partial^2 x^0}{\partial s^2}, \frac{\partial^2 \mathbf{r}}{\partial s^2}\right\}.$$

It is perpendicular to the 4-velocity because its scalar product with the velocity is

$$\frac{\partial(\partial \mathbf{s}/\partial s)}{\partial s}\cdot\frac{\partial \mathbf{s}}{\partial s},$$

or

$$\frac{\partial(\partial \mathbf{s}/\partial s)^2}{\partial s},$$

which is zero because the magnitude of $(\partial \mathbf{s}/\partial s)$ is a constant.

Since the 4-velocity has a constant magnitude, it might correspond to some physical quantity; that can be elucidated by observing that momentum of a particle in 3-space is the 3-velocity multiplied by a mass, $\mathbf{p} = m\mathbf{v}$.

Thus if we multiply the space-like components of the 4-velocity by a mass we obtain the components of the momentum as the space-like components of a new 4-vector that has the components

$$\gamma m(c, \mathbf{v}),$$

exhibiting the speed of light explicitly.

The square of the magnitude of the vector is m^2c^2. It must be a constant since it is the square of the magnitude of a 4-vector and it is clear that in the local frame of rest it can only be proportional to the square of the conventional mass.

If we multiply the vector by the speed of light we obtain a vector that has the dimensions of energy and which has a constant magnitude, mc^2; let us denote it by E_0. The three space-like components of the vector are the components of the momentum multiplied by c, and so have the dimensions of energy, while the time-like term also has the dimensions of energy. If we denote the time-like component by E, the equation for the magnitude of the vector reads

$$E^2 - c^2p^2 = E_0^2.$$

To interpret the relation, let E be put equal to $E_0 + \epsilon$, where in the non-relativistic condition when cp is small, the difference ϵ is small compared to E_0.

Then

$$2E_0\epsilon = c^2p^2,$$

and so

$$\epsilon = c^2p^2/2E_0 = p^2/2m.$$

Thus the difference between E and E_0 is just the classical kinetic energy for small velocities.

E_0 is conventionally called the rest energy of a particle but it is more correctly to be thought of as the mass of the particle as measured in the local laboratory. It is not possible to make comparisons between the local laboratory standard of mass and another in a remote laboratory so that the observable quantity that has to be kept constant in any transformation of coordinates is the mass as measured locally. That we call the *rest mass*, m_0 from now on, and the constant energy obtained by multiplying it by c^2 we call the *rest energy*.

A somewhat different approach to energy and momentum is as follows. Two functions of velocity arise in non-relativistic mechanics, the momentum which is a linear function of velocity and is a 3-vector, and kinetic energy which is a scalar and a quadratic function of velocity. We seek a 4-vector that has the 3-momentum as three of its components and that has a constant magnitude under a Lorentz transformation so that it may correspond to some measured physical quantity. The Lorentz transformation is a linear transformation so that if a vector is to transform correctly under it, the components of the vector must be homogeneous: we cannot combine the momentum and classical kinetic energy as components of a new vector, for such an inhomogeneous object would not transform properly under the Lorentz transformation. If we are to construct a 4-vector linearly related to momentum we must associate with the three components of momentum a fourth linear term. Since the momentum is proportional to the first power of the mass, the additional term must also be proportional to the first powers of the mass and of a velocity. The factor of proportionality cannot however be a function of the 3-velocities, since they are already used in the 3-momentum and the only possibility that keeps the correct dimensions is to multiply the mass by the speed of light. Thus we obtain a 4-vector with components

$$(mc, m\mathbf{v}).$$

But that is just proportional to the 4-velocity and for it to have an invariant magnitude the mass (which is the classical mass m_0 at small velocities) must be set equal to γm_0.

This discussion must seem rather remote from the previous arguments because we have said nothing about what observations of momentum and energy can be made at a distance nor indeed how the mass of a distant body can be measured. We have generated a consistent mathematical scheme which has a formal relation with physics but it is not clear at this stage how distant observations of energy and momentum would be made, nor whether it is meaningful to ascribe an energy to a particle that is locally at rest. We do know that composite particles at rest have energies that can be transformed into, for example, electromagnetic radiation, but that rather begs the question of whether truly elementary particles, whatever they may be, have rest energies or whether the rest energy of such a particle may not be just a book-keeping number.

Ways of relating energy and momentum at a distance to the local standards of mass (at Sèvres) can now be envisaged, but to explain them we must first look at the electrical quantities.

3.4 Electromagnetism

It was seen in Chapter 2 that electromagnetic quantities are nowadays related to frequency through the Josephson effect and the quantum Hall effect. How may we determine a distant electromagnetic field from the observed frequency of radiation applied to a Josephson junction?

Let a coil of area A be rotated at an angular frequency of rotation ω in a stationary field $\mathbf{B}$. (Figure 3.3). Then the voltage induced between the ends of the coil oscillates with the amplitude $BA\omega$ at the frequency ω. By some means we detect the occurrence of steps in the current–voltage characteristic of a Josephson junction to which the induced voltage is applied. If ν is the frequency of an applied radio frequency field, we shall have

$$BA\omega = (h/e)\,\nu.$$

Now the frequency ω is that of the periodic variation of the induced voltage and the value that the observer records is changed by the relativistic Doppler shift. The observer also makes an independent measurement of the frequency ν as received at her position, and that too is changed by the relativistic Doppler shift. Since the relative shift is independent of frequency, the ratio ν/ω, equal to ρ, say, is independent of the velocity of the source relative to the observer. Consequently, the combination

$$BA = \rho(h/e),$$

is an invariant of the observations.

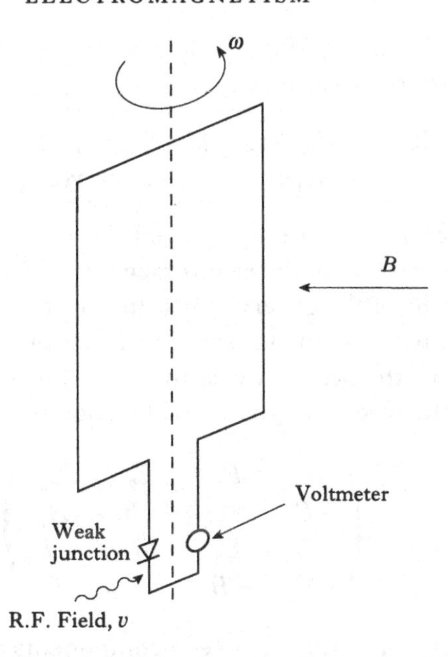

Figure 3.3. EMF generated by coil rotating in magnetic field.

The nature of the argument needs a little development. The whole aim of this chapter has been to identify quantities that are the results of observation and thus invariant. We then have to set up some mathematical model from which those invariant quantities may be constructed and we have proceeded by analogy with what we do locally. We do the same here. Local magnetic effects can be accounted for in terms of a magnetic field vector derived from a vector potential and we assume that the same rules apply at the remote experiment. It would indeed be a strange physics if we had to change the form of a theory in a distant region when we have no direct knowledge of the empirical basis of the theory in regions remote from us. We must suppose that it is possible to define a potential and its derivatives in a distant region just as it is in our own locality. The consequence is that the scalar product of field and area must be an observable invariant.

Area in three-dimensional space is the vector product of two linear vectors $\mathbf{x}$ and $\mathbf{y}$; it is therefore an axial vector and corresponds to an antisymmetrical tensor of second rank with components $A_{ij} = (\mathbf{x} \wedge \mathbf{y})_{ij}$. It is the three-dimensional part of a second rank tensor in 4-space which transforms as follows for a velocity v ($c = 1$) in the 1-direction:

$$A_{02} = \gamma(A_{02}+vA_{12}), \quad A_{03} = \gamma(A_{03}+vA_{13});$$
$$A_{12} = \gamma(A_{12}+vA_{02}), \quad A_{13} = \gamma(A_{13}+vA_{03});$$

all other components are unchanged.

Since the product of the electromagnetic field and area is a scalar and a constant of the observations, the electromagnetic field must be represented by an antisymmetric second rank tensor that transforms in the same way as the area. Formally, therefore, we may write the electromagnetic field tensor as

$$\begin{pmatrix} 0 & E_1 & E_2 & E_3 \\ -E_1 & 0 & -B_3 & B_2 \\ -E_2 & B_3 & 0 & -B_1 \\ -E_3 & -B_2 & B_1 & 0 \end{pmatrix},$$

with the transformations between components as given in Section 3.1

In local theory the components of that tensor are written as the space-time derivatives of a four-dimensional potential (A_0, A_1, A_2, A_3), or ($A_0, \mathbf{A}$), where $\mathbf{A}$ is the local vector potential and A_0 the scalar potential, in particular, locally, $\mathbf{B} = \mathbf{curl A}$. The derivative operator is a 4-vector and it therefore follows that if the electromagnetic field tensor is to transform as a tensor of rank 2, then the potential ($A_0, \mathbf{A}$) must also transform as a 4-vector.

In this argument, no experiment has been suggested that would involve the electric field directly, and it seems difficult to devise one that would give an invariant result, for all observable quantities appear to involve a time or frequency which, in the absence of other information, cannot be related to the time or frequency at the distant site.

We saw at the end of Chapter 2 that it is in principle possible to abandon the metal standard of mass and replace it by a standard of force or work based on the relation of voltage to frequency, together with the quantum Hall resistance. Since the ratio (h/e) and the quantum Hall resistance are scalars and by definition universal constants, the local standard of mass may be transferred to a distant laboratory using measurements of Josephson frequencies as just discussed. Rest mass itself is a scalar and its value must be independent of the frame in which it is observed but whereas in the past it was not possible to assign values to masses at distant places, that can now be done in terms of our local laboratory standards through electromagnetic signals and scalar universal constants.

3.5 General relativity

One of the axioms of special relativity is that the speed of light must be supposed constant because we have no means of measuring it at a distance, and in consequence we argued in Section 3.2 that light times must be combined according to the rules of Euclidean geometry and that light paths must be taken to be straight lines. If the speed of light is a constant independent of position, the coefficient g_{00} in the metric coform $g_{ik}\,\mathrm{d}x^i\,\mathrm{d}x^k$ must be a constant; it is 1 if the speed of light is taken to be 1. The coefficients g_{0i} are likewise zero.

Einstein argued that the speed of light was not necessarily the same everywhere and that a more general form of metric should be used, with the metric coefficients dependent on the distribution of matter, and he derived the well-known consequences, the gravitational red shift, the deflection of light by the Sun and the advance of the perihelion of Mercury (now far better determined by radar observations, Shapiro, 1980) that for many years remained the only observational support for general relativity. We may now restate the argument in conformity with the approach of this chapter, for in recent years we have had cogent direct observational evidence that is inconsistent with the hypothesis that the speed of light is independent of position. The optical observations of the deflection of light by the Sun were difficult to make and afflicted by serious systematic errors, though they did show fairly convincingly that the deflection existed and was consistent with general relativity and not with Newtonian gravitational theory. Radio observations now afford much more compelling evidence that the speed of light varies with position in the neighbourhood of the Sun, for the angular deflection of a ray has been measured with high precision by radio interferometers (Fomalmont and Sramek, 1977) and the time delay of a signal along a ray passing close to the Sun has been measured directly (Shapiro *et al.*, 1972, Anderson *et al.*, 1975,

Reasenberg *et al.*, 1979). The implication of those observations is that the value of the square of a space-time interval, ds^2 depends on the direction in which the events are observed, how closely the ray paths from the events to the observer pass by great masses.

Special relativity remains the appropriate space-time geometry for 'local' observations and for regions far from great masses but clearly in the neighbourhood of the Sun, the speed of light is not independent of position. The more general metric form of general relativity has to be used for paths close to the Sun, with coefficients that become those of special relativity in regions remote from large masses. The choice of the metric coefficients is made by fitting them to observation, in particular, to ensure that the variation of the speed of light close to the Sun is consistent with the dynamical effects on the orbits of Mercury, Venus and the Earth. The proper choice is the Schwarzschild metric:

$$ds^2 = \left(1+\frac{2U}{c^2}\right)(dx^0)^2 - \left(1-\frac{2U}{c^2}\right)\{(dx^1)^2+(dx^2)^2+(dx^3)^2\},$$

where U, the gravitational potential is $-GM/r$ at a distance r from a body of mass M (see Appendix, where different sign conventions are used in Resolution A4 of the General Assembly of the International Astronomical Union, 1991).

The dynamical consequences of general relativity depend upon a further principle enunciated by Einstein, the principle of equivalence, that the paths of test bodies, effectively point objects, are geodesics in space-time, the same for all such objects irrespective of constitution. It is implicit in the calculations that relate the change of the speed of light to the precessions of planetary orbits and it also leads to the important conclusion that gravitational forces observed in the laboratory are independent of constitution, a conclusion that has been verified to within about 1 part in 10^{12} (see Chen and Cook, 1993).

The identification of the quantity U in the Schwarzschild metric with the Newtonian potential $-GM/r$, depends upon the equivalence principle of Einstein, for when the ratio U/c^2 is small, Einstein's equations of motion reduce to those of Newton. Thus the solar value of GM must be equal to the Keplerian product, $\omega^2 a^3$, where ω is the mean motion of a planet and a is the major semi-axis of its orbit. The solar value is very well known, but the mass, M, can only be calculated in laboratory units if the value of G has been found experimentally in a terrestrial laboratory. That value is very uncertain (Chen and Cook, 1993).

General relativity is now so accepted that it is explicitly incorporated into fundamental astronomical definitions, as set out in the Appendix. It has to be allowed for in comparing times and frequencies to the highest precision, for different frequency standards operate in different gravitational potentials. In addition, the conventional, internationally adopted value of the speed of light is applicable only in free space, not only free of actual matter but also where the potential of adjacent masses is effectively that of the vacuum.

3.6 Conclusion

The argument of this chapter has been based on the distinction between observed quantities and mathematical objects such as coordinates, electric fields and so on, that constitute the framework of theoretical models. Observed quantities can never be altered, the framework of theory can be varied at our convenience, but always subject to the condition that the calculated results of observations coincide with the actual unchangeable measurements. Geometry and electrodynamics of distant events are models, the only observations we can make are light times, frequencies and other properties of electromagnetic signals, and those must remain invariant as we change the geometry and electrodynamics.

So far we have considered events that are not directly accessible because they are at a great distance. There are also events that are not directly accessible because they are on too small a scale and of which the only knowledge we have comes through electromagnetic radiation emitted or absorbed when a microscopic system undergoes some change. They are the subject of the next chapter.

4

Microphysics: relativistic quantum mechanics

4.1 Introduction

IN THE TWO PREVIOUS CHAPTERS we have seen how the ways in which we establish standards of time and make observations at a distance entail the equation of time evolution of quantum mechanics and the space-time geometry of special relativity. Quantum mechanics and special relativity are often thought of as separate extensions of classical mechanics, on the one hand when Planck's constant cannot be neglected and on the other, when the speed of light is treated as infinite. It should however be clear from the arguments of Chapters 2 and 3 that the ways in which we establish our standards of measurement and the sorts of observations that we can make, imply that our theories of physical systems should accord with both quantum mechanics and, whenever direct measurements of distance cannot be made, with special relativity. The interior of the atom is inaccessible to direct measurements of distance, no less than very distant objects and so we must ask whether the observations that we can make of microscopic systems entail a relativistic form for theories of atomic structure. If we model internal structures by interacting particles, the velocities of those particles are relativistic so that theories of microscopic physics should be in Lorentz-invariant forms.

The formalism of Chapters 2 and 3 is accordingly extended to microscopic systems. The observations that can be made of microscopic systems are quite different from those of macroscopic systems. We accept that atoms and molecules are made up of more elementary particles, but we can make no direct observations of the relations between them, of their relative positions and velocities or the electromagnetic fields that they establish or to which they are subject. Atoms and molecules are bound systems and all we can observe are their interactions as entities with the external world and that only when they change their states. Thus we can observe the frequency, polarisation and intensity of electromagnetic radiation that they emit or absorb. We can also observe the path of a bound system in an electromagnetic field, as was seen in Chapter 2 where the deflection of atoms in a non-uniform magnetic field was used to distinguish between atoms of caesium in different states. Formal theories of bound microscopic systems postulate internal structures while being constructed to satisfy the observations, to be consistent with the ideas we already have about quantum mechanics and special relativity, and to have a capacity to predict.

We have in each of the two previous chapters seen that there are two ways of establishing theoretical models, on the one hand making use of *a priori* arguments and then checking the results against observation, and on the other hand, following the implications of the nature of possible observations. In the same way, in microphysics, we may on the one hand obtain the form of quantum mechanics in the geometry of special relativity and then consider how far it leads to the observations. Alternatively, we may start with observations and see if they necessarily lead to some formal structure. The distinction is one of logic and does not necessarily correspond to the historical development.

The metrological requirement upon which special relativity depends is that certain combinations of light times should be invariant under transformations of space-time coordinates. In large scale physics we are concerned with the time it takes for information to be transmitted by electromagnetic waves, the time that is, for a change in an electromagnetic field, corresponding to information about distant events, to travel from place to place. In microphysics, in just the same way, we assume that changes in electromagnetic fields take time to travel from place to place, but now those changes correspond to forces on charged particles. We are concerned with the time it takes for a force to be exerted on one particle as a result of the motion of another charged particle. That time delay has to be incorporated into the quantum mechanics of charged particles in electromagnetic fields.

We emphasise again, in line with the argument of previous chapters, that the quantum mechanics of microsystems is a model, that our observations of microsystems are of the results of changes in such systems and that we cannot observe the state of a microsystem directly.

4.2 Quantum mechanics in the geometry of special relativity

We have seen in Chapter 3 that when information about distant events is transmitted by electromagnetic radiation, times and distances of the events are correctly related by the Lorentz transformation, and also that the equations of electromagnetism are themselves invariant under the transformation. In so far as in our models of atomic structure, the forces between microscopic particles are supposed to be electromagnetic, it follows that times and distances in those models of microphysics should transform according to the Lorentz transformation. We also saw in the previous chapter that the properties of the 4-velocity entailed a relation between energy and momentum corresponding to the invariance of the magnitude of the 4-vector formed from the energy and the components of the momentum:

$$(E, c\mathbf{p})\cdot(E, c\mathbf{p}) = E_0^2,$$

where E_0 is the rest energy, $m_0 c^2$.

Energy and momentum in the formalism of quantum mechanics are operators upon the wavefunction and correspond to the differential operators, $\partial/\partial t$ and $\partial/\partial \mathbf{r}$ respectively, where $\partial/\partial \mathbf{r}$ is the vector of spatial derivative operators. Thus, putting arbitrary constants equal to 1, the energy–momentum relation of special relativity translates into

$$\frac{\partial^2 \Psi}{\partial t^2} - \frac{\partial}{\partial \mathbf{r}} \cdot \frac{\partial}{\partial \mathbf{r}} \Psi = E_0^2 \Psi.$$

That is the Klein–Gordon equation for a particle of rest energy E_0. It is evidently invariant under the Lorentz transformation but it is unsatisfactory in two respects. It is not of the form of the quantum mechanical equation of time evolution, in which the derivative with respect to time should be of first order; and objects to which it applies are not conserved. If the wavefunction is to contain all the information about the state of a system, it must satisfy a first order differential equation for evolution in time, for otherwise more than a single initial value would be needed to determine the wavefunction at later times (Chapter 2). We also saw that the same form of equation is entailed by the operational definition of the standard of frequency. An equation that involves the second differential, as the Klein–Gordon equation does, is incompatible with that condition and the wavefunction cannot contain all the information about a system.

Solutions of the Klein–Gordon equation cannot represent particles that are conserved because they are indeterminate in the sense that they depend on conditions at more than one initial time. The following argument demonstrates the point explicitly. Pre-multiply the Klein–Gordon equation by the complex conjugate of the wavefunction and post-multiply the complex conjugate of the equation by the wavefunction:

$$\Psi^*\frac{\partial^2\Psi}{\partial t^2}-\Psi^*\frac{\partial}{\partial \mathbf{r}}\cdot\frac{\partial}{\partial \mathbf{r}}\Psi=\Psi^*E_0^2\Psi,$$

$$\left\{\frac{\partial^2\Psi^*}{\partial t^2}\right\}\Psi-\left\{\Psi^*\frac{\partial}{\partial \mathbf{r}}\cdot\frac{\partial}{\partial \mathbf{r}}\right\}\Psi=\Psi^*E_0^2\Psi.$$

If the second equation is subtracted from the first, the result is

$$\frac{\partial}{\partial t}\left\{\Psi^*\frac{\partial\Psi}{\partial t}-\frac{\partial\Psi^*}{\partial t}\Psi\right\}=\frac{\partial}{\partial \mathbf{r}}\left\{\Psi^*\frac{\partial\Psi}{\partial \mathbf{r}}-\frac{\partial\Psi^*}{\partial \mathbf{r}}\Psi\right\}.$$

The left hand side is not positive definite and so the equation tells us nothing about the time evolution of the particle number or probability, $\langle\Psi^*\,|\,\Psi\rangle$.

When the same operations are performed on the time-evolution equation of quantum mechanics in the form

$$\mathrm{i}\partial\Psi/\partial t=H\Psi,$$

remembering that H is Hermitian, the result is

$$\mathrm{i}\left\{\left\langle \Psi^* \middle| \frac{\partial \Psi}{\partial t}\right\rangle + \left\langle \frac{\partial \Psi^*}{\partial t} \middle| \Psi \right\rangle\right\} = \langle \Psi^* | (H\Psi)\rangle - \langle (\Psi^* H) | \Psi \rangle,$$

or

$$\mathrm{i}\frac{\partial \langle \Psi^* | \Psi \rangle}{\partial t} = \langle \Psi^* | (H\Psi)\rangle - \langle (\Psi^* H) | \Psi \rangle.$$

The right side evidently vanishes for a stationary state for which $H\Psi = E\Psi$. It also vanishes if H is equal to the Laplacian plus a real potential energy operator.

Consequently the square of the norm of the wavefunction is conserved in those circumstances and can stand for the number or probability of particles. If the number of particles is given initially, it is determined for all time.

For those reasons, the time evolution equation of quantum mechanics for conserved particles should have a time derivative of first order only. Now if the whole equation is to be invariant under the Lorentz transformation, it must be homogeneous, which means that the spatial derivatives of the wavefunction must also be of first order, that is that the Hamiltonian should be a linear combination of the momentum operators themselves and not of their squares as in the classical form of the kinetic energy. That may seem arbitrary, or inconsistent with classical mechanics, but the form of the classical Hamiltonian depends on macroscopic experiments done at small velocities and the results are therefore not necessarily applicable to relativistic conditions or to the internal state of a microscopic bound system. The classical equations of motion involve an Hamiltonian function and the quantum mechanical equation of time evolution involves an Hamiltonian operator, but nothing in the general theory tells us what form that function or that operator should take, and in fact they must be chosen to reproduce the observed empirical physics.

The classical relativistic relation between energy and momentum is

$$(E, c\mathbf{p})\cdot(E, c\mathbf{p}) = E_0^2$$

in which the rest energy appears as the scalar product of the energy–momentum 4-vector with itself; in any modification of that relation we must retain the invariant rest energy. If we are to have a linear combination of the energy and momentum operators, we must seek a scalar product of the energy–momentum 4-vector with some 4-vector operator, which must be a constant because it cannot depend on the energy and momentum. Let us denote such a vector by $(\gamma^0, \boldsymbol{\gamma})$, where $\boldsymbol{\gamma}$ is a 3-vector. We may take the square of its magnitude $\{(\gamma^0)^2 - \boldsymbol{\gamma}\cdot\boldsymbol{\gamma}\}$, to be 4.

We then require that the operation by the scalar product, $(\gamma^0, \boldsymbol{\gamma})\cdot(E, c\mathbf{p})$, namely $(\gamma^0 E - c\gamma^i p_i)$, should multiply the wavefunction by a constant.

Repeated operations by the scalar product give

$$(\gamma^0, \boldsymbol{\gamma})\cdot(E, c\mathbf{p})\,(\gamma^0, \boldsymbol{\gamma})\cdot(E, c\mathbf{p})\,\Psi,$$

which should also have the effect of multiplying the wavefunction by a constant.

We determine the elements of the constant vector from the requirement that two successive operations by the scalar product $(\gamma^0 E - c\gamma^i p_i)$ should multiply a wave function by E_0^2, that is to say,

$$(\gamma^0 E - c\gamma^i p_i)\cdot(\gamma^0 E - c\gamma^i p_i)\,\Psi = E_0^2\,\Psi.$$

The operator must accordingly be equivalent to the classical expression

$$E^2 - c^2 p^2 (= E_0^2),$$

where p^2 is $\mathbf{p}\cdot\mathbf{p}$.

We then obtain the following commutation relations:

$$(\gamma^0)^2\cdot\Psi = \Psi, \quad (\gamma^i)^2\cdot\Psi = -\Psi,$$
$$\{\gamma^0\gamma^i - \gamma^i\gamma^0\}\,\Psi = 0,$$
$$\{\gamma^i\gamma^k - \gamma^k\gamma^i\}\,\Psi = 0.$$

Those relations cannot be satisfied by ordinary numbers and we must take the elements to be anti-commuting operators. As is well known there is a number of possible forms; here we take them to be the following elements of a Clifford algebra (Penrose, 1986):

$$\gamma^0 = \begin{pmatrix} I & 0 \\ 0 & -I \end{pmatrix}, \quad \gamma^i = \begin{pmatrix} 0 & \sigma_i \\ -\sigma_i & 0 \end{pmatrix},$$

where I is the 2×2 unit diagonal matrix and σ_i is a 2×2 Pauli matrix.

The time-evolution equation thus becomes

$$i\gamma^0 \frac{\partial \Psi}{\partial t} = E_0 \Psi + c\gamma^i p_i \Psi,$$

or, if we multiply throughout by γ^0,

$$i\frac{\partial \Psi}{\partial t} = \gamma^0 E_0 \Psi + c\gamma^0 \gamma^i p_i \Psi.$$

Planck's constant has been put equal to unity.

The second form is in the conventional form of quantum mechanics with an Hamiltonian equal to

$$\gamma^0 E_0 + c\gamma^0 \gamma^i p_i.$$

However in that form the time evolution equation is not invariant under Lorentz transformations and to obtain an invariant form, the whole equation must be operated on by γ^0.

The energy of an eigenstate is given by

$$(\gamma^0 E_0 + c\gamma^0 \gamma^i p_i)\Psi = E\Psi$$

and on operating twice with the Hamiltonian we have

$$(E_0^2 + c^2 p^2)\Psi = E^2 \Psi.$$

The Hamiltonian just obtained is for a free particle. It is well known that it does not commute with the classical angular momentum $\mathbf{r} \wedge \mathbf{p}$, but that it does commute with $(\mathbf{r} \wedge \mathbf{p} + \frac{1}{2}\boldsymbol{\sigma})$, where $\boldsymbol{\sigma}$ is the 4×4 operator diag$[\sigma_i, \sigma_i]$, in which σ_i is again a 2×2 Pauli matrix.

The Hamiltonian for a charged particle in an electromagnetic field is supposed to be obtained, as in classical electrodynamics, by adding the 4-vector of the electromagnetic potentials to the energy–momentum 4-vector; thus

$$\mathbf{p} \Rightarrow \mathbf{p} + e\mathbf{A},$$
$$E \Rightarrow E - eA_0.$$

It is then found that a charged particle of mass m in a magnetic field behaves as if it had an intrinsic magnetic moment $\mu = eh/m$.

The spin–orbit interaction in atoms and the spin–axis interaction in molecules are further consequences of that Hamiltonian, and the structures of atomic energy levels follow from the properties of the symmetry group of an Hamiltonian in which the potential is that of a central field. Likewise, the structures of the electronic energy levels of a molecule follow from the properties of the symmetry group when the electric field has the symmetry of the ionic framework of the molecule.

So far, we have followed the usual exposition, explaining how it is possible to construct a quantum mechanical Hamiltonian consistent with special relativity, and indicating the consequences of that Hamiltonian for the behaviour of single particles and atomic and molecular structure, consequences which are commonly stated in terms of the spins of elementary particles. Let us now start from, rather than end with, the observations.

4.3 From observation to theory

When atoms are placed in a uniform magnetic field, we often observe that a single state in which they are in the absence of the field, appears to be replaced by a number of states that depend on the magnitude of the applied field. We have already met an example in the experiment used to determine the standard of frequency (Chapter 2), for the separation of atoms into two beams is brought about by passing them through a magnetic field, a field that has to be non-uniform if forces are to be exerted on atoms in the two states. Here atoms in the two states are physically separated by the different forces acting on them.

Experiments on leptons, electrons and muons, show that they have two and only two states or polarisations in a magnetic field, and they are found to precess or change their states when subject to electromagnetic torques.

More commonly we observe radiation emitted or absorbed as atoms change from one state to another. Again atomic frequency standards provide examples. A signal at the standard frequency brings about a change of state and in the absence of a magnetic field in the region where the radio frequency field is maintained, there is just a single transition frequency. If however, a steady magnetic field is maintained in that region it is found that transitions can be brought about by fields at any of a number of frequencies. (In the practical operation of the standard a very weak field is maintained to ensure that the desired frequency is distinct from all the others.)

Optical transitions likewise occur between atoms in different states, only it is usually not possible to distinguish those in different states as is done in the frequency standard; instead, guided by our knowledge of what happens in the standard, we postulate that optical transitions occur between atoms in different states. Historically of course, the notion that optical transitions took place between atoms in different states arose long before an atomic beam frequency standard was developed, but the sequence just set out is the more logical. When a magnetic field is applied, a single optical transition is replaced by a number of transitions with frequencies close to the original frequency, the differences being closely proportional to the strength of the magnetic field. The radiation in the absence of the field is unpolarised, but the transitions in the presence of the field are polarised; the sense of the polarisation and whether it is linear or circular depends on the direction of the radiation relative to that of the field. It is not difficult to work out from the number of transitions and their polarisations the number of substates into which each principal state goes when a magnetic field is applied.

In the absence of a field, there is no preferred direction in space relative to an observer and atoms have arbitrary orientations. Since they emit or absorb radiation at a single frequency, the state of the atom is independent of direction and is completely spherically symmetrical. When the field is applied a preferred direction is established in space and within the atoms and it now appears that the atoms can exist in a number of states instead of just one. The number of states to which the original single state gives rise may be even in some cases and odd in others.

We are concerned here with the behaviour of leptons and atoms under symmetry operations, that is to say, arbitrary rotations or rotations around a given axis, and the natural language in which to discuss those matters is that of groups and their representations. An abstract group may be represented by a set of operators acting on basis functions, in other words, we are concerned with groups acting on vector spaces. Commonly of course, operators are thought to be mathematical operators and the basis to be a set of mathematical functions or vectors in some appropriate space. However, that is not necessary, and the reason that group theory is so fertile in its application to physics is that groups may also be represented by, and be isomorphic to, sets of physical operations that act on physical systems. That will be taken much further in Chapter 6; here it is enough to note that in observing a gas of atoms in the absence or presence of a magnetic field, we are looking at representations of the rotation group in three dimensions or of a sub-group of it; that is because atoms in a gas have random orientations and it follows that the frequency of any transition that we observe, being the mean of frequencies observed in all possible directions, is independent of orientation; note that we observe the radiation from a population of atoms and it is the population that is spherically or otherwise symmetrical not an individual atom, which we do not observe.

We now see how to establish a mathematical model corresponding to the physical representation of the rotation group in 3-space, just as in Chapter 3 we found a model that gave the correct invariance under transformations of space-time coordinates. Consider first states of an atom that are invariant under rotation. The following argument looks as though it applies to an individual atom, in conflict with what has just been said about observing populations. However, since the atoms in a gas may be considered to be independent for this purpose (ignoring, for example, pressure broadening of a transition), the overall wavefunction of the population is the product of individual wavefunctions and the argument below applies to each of the latter separately.

Let the eigenvalue equation for any such state invariant under rotation be

$$H_0 \Psi = E_0 \Psi,$$

where Ψ is an eigenfunction of H_0; H_0 is by definition an operator invariant under rotation.

Let R be any rotation in 3-space. Since the eigenenergy is unaffected by rotation, we have

$$R(H_0 \Psi) = R(E_0 \Psi) = E_0 (R\Psi),$$

or

$$RH_0 R^{-1} R(\Psi) = E_0(R\Psi).$$

Now $RH_0 R^{-1}$ is equal to H_0 because H_0 is invariant under rotation and therefore

$$H_0(R\Psi) = E_0(R\Psi),$$

that is, $R\Psi$ is also an eigenfunction of H_0. But $R\Psi$ is an arbitrary linear combination of a set of eigenfunctions of H_0 and it follows that every such eigenfunction of H_0 has the same eigenenergy.

The number of eigenfunctions with the same eigenenergy is found from the behaviour of the atom in a magnetic field. When a uniform field is applied to an atom, the Hamiltonian (H) becomes (H_0+H_B), where H_B is approximately proportional to the field, at least for weak fields. The component H_B is not invariant under rotations about an arbitrary axis and so neither is H, but H_B is invariant under any rotation, A, about the direction of the field. Thus we have

$$AHA^{-1}A\Psi_i = E_i A\Psi_i,$$

for some eigenfunction, Ψ_i, of H.

Provided H_B is small, perturbation procedures allow us to express Ψ_i as a linear combination of eigenfunctions of H_0 and in that way the eigenfunctions of H_0 can be classified into sets in which each has the same eigenenergy of H and transform into each other under the axial rotation A. The number of eigenenergies of H is equal to the number of sets of eigenfunctions that are equivalent under A.

Let us try to place those facts in terms of the symmetry of the atom under rotations in 3-space. In the absence of a magnetic field, the atom is in a state that is completely symmetrical with respect to arbitrary rotations in 3-space. When a uniform magnetic field is applied, a preferred direction is established, that of the field, and the symmetry is reduced from complete spherical symmetry to axial symmetry about the direction of the field. In the language of group theory, the states of the atom form a basis for the representation of the group of rotations by sets of rotation operators acting upon vectors in 3-space. Any symmetry operator S has the property that when it operates on a wavefunction it transforms the wavefunction into a linear combination of wavefunctions that have the same eigenenergy:

$$S\Psi_i = a_{ij}\Psi_j,$$

where

$$H\Psi_j = E\Psi_j \quad \text{for all } j.$$

Hence

$$HS\Psi_i = Ha_{ij}\Phi_j = Ea_{ij}\Phi_j,$$

while

$$SH\Upsilon_i = SE\Upsilon_i = Ea_{ij}\Phi_j,$$

so that symmetry operators commute with the Hamiltonian.

Sets of spherical harmonics are obvious bases for representations of the rotation group. The radial distance, r, from a centre is invariant under any rotation, while the harmonics of first order, namely

$$r^{-2}(x/r, y/r, z/r),$$

transform into each other under arbitrary rotations. In complete spherical symmetry the directions of the coordinate axes are arbitrary and all three wavefunctions would have the same eigenenergy. If there were symmetry about an axis, as when a uniform magnetic field is applied to an atom, two of the coordinate directions, x and y say, would be arbitrary and the eigenenergies corresponding to them would be the same, whereas the eigenenergy corresponding to the function z/r^3 would be different, and moreover would depend on the direction of the field, that is, on the sign of z. Consequently, in the presence of a magnetic field, an atom would have three eigenenergies, one independent of the field and two, of opposite sign, proportional to the field. We see from this argument how we may represent the state of an atom that has one eigenenergy in the absence of a magnetic field and three when a field is applied – the wavefunctions that are the basis of the representation must correspond to coordinates in three dimensions and the symmetry operators will be the 3×3 rotation matrices.

The group of rotations in three dimensions and the Lorentz group in four dimensions are examples of Lie groups. A Lie group is a group in which the elements depend upon the values of one or more parameters, such as the angles of rotation, that can take any values within a specified range, 0 to 2π for angles. A tangent operator can be defined as the rate of change of the operator with respect to the parameters at the origin of the parameters. When the parameters are all zero, the element of the group is just the identity operator, but the tangent operator is not zero. It will however be independent of the parameters for linear transformations, and in fact the whole group structure is determined by the algebra of the tangent operators at the origin.

Consider a representation of rotations in 3-space. Let $\delta\theta$ be an arbitrary rotation – it is a vector. Let $\mathbf{v}$ be a vector and let the rotation transform it into $\mathbf{v}'$. Then we may write

$$(1+\mathbf{M}\cdot\delta\boldsymbol{\theta})\,\mathbf{v} = \mathbf{v}' = \mathbf{v}+\delta\mathbf{v},$$

whence $\mathbf{M}$ is the derivative,

$$\partial\mathbf{v}/\partial\boldsymbol{\theta} \text{ at } \delta\boldsymbol{\theta} = 0.$$

$\mathbf{M}$ is a vector operator with components, M_i, that are matrices. It is the angular momentum operator.

In ordinary 3-space the magnitude of the vector $\mathbf{v}$ is a measured quantity and the metrological condition on the transformation of the components of $\mathbf{v}$, namely that the magnitude of $\mathbf{v}$ should be unaltered, is therefore

$$\mathbf{v}'^T\cdot\mathbf{v}' = \mathbf{v}^T\cdot\mathbf{v},$$

when the products operate on the wavefunction ($\mathbf{v}^T$ denotes the transpose of $\mathbf{v}$).

Thus

$$\mathbf{v}^T(1+\delta\boldsymbol{\theta}^T\cdot\mathbf{M}^T)\cdot(1+\mathbf{M}\cdot\delta\boldsymbol{\theta})\,\mathbf{v} = \mathbf{v}^T\mathbf{v},$$

so that

$$\mathbf{v}^T\delta\boldsymbol{\theta}^T\cdot\mathbf{M}^T\mathbf{v}+\mathbf{v}^T\mathbf{M}\cdot\delta\boldsymbol{\theta}\mathbf{v} = 0.$$

Consequently $\mathbf{M}$ is antisymmetric. In fact the components of $\mathbf{M}$ are

$$\begin{pmatrix} 0 & +1 & 0 \\ -1 & 0 & 0 \\ 0 & 0 & 0 \end{pmatrix}$$

and two similar matrices.

The commutation relations for angular momentum, namely

$$M_i M_j - M_j M_i = M_k,$$

follow from the form of M_i.

According to Lie group theory, the operator for a finite rotation $\boldsymbol{\theta}$ is $\exp(\mathbf{M}\cdot\boldsymbol{\theta})$.

In the description of the properties of a Lie group, we have derived the commutation rule for the angular momentum operator from a particular representation of the rotation operator in 3-space and the constancy of the magnitude of a vector under rotation, but the rule is a quite general consequence of the conservation of angular momentum, itself a consequence of the invariance of physics with respect to the origin of angular coordinates (see Chapter 5).

Since the angular momentum operators operate on basis functions of a representation of the rotation group, their dimensions are determined by the dimensions of the representation, while the number of elements of the algebra is equal to the number of independent operators of those dimensions. Thus, if the dimension of the representation is 3, as it is when the basis functions are coordinates in 3-space, it is easily seen that there are just three independent operators (in addition to the unit operator) that satisfy the commutation conditions. There must in fact be just three such operators in 3-space. There are also three independent operators, the Pauli spin matrices, when the dimension of the representation is 2 and the basis functions are spinors.

The 3×3 matrices represent rotations in real 3-space but they can easily be combined with a matrix representing a boost. Consider a rotation in the z-direction and a boost in the same direction. The system as a whole is symmetrical about the z-axis and the matrices for the transformation of the four coordinates of space and time are (time first):

$$\begin{pmatrix} 1 & . & . & . \\ . & \cos\theta & \sin\theta & . \\ . & -\sin\theta & \cos\theta & . \\ . & . & . & 1 \end{pmatrix} \text{ and } \begin{pmatrix} \cosh\zeta & . & . & \sinh\zeta \\ . & 1 & . & . \\ . & . & 1 & . \\ \sinh\zeta & . & . & \cosh\zeta \end{pmatrix}.$$

The angle of rotation in 3-space is θ and $\tanh\zeta$ is equal to v/c, where v is the boost velocity.

The two matrices are independent and the product, the matrix corresponding to a rotation and a boost, partitions into two independent 2×2 diagonal matrices. Thus the representation accommodates rotations and boosts and the 3×3 rotation matrix can easily be seen to be a principal diagonal submatrix of the complete 4×4 matrix.

While a representation with three basis functions is the natural one for rotations in three dimensions, it is not the only possible one and in fact cannot accommodate a situation in which a state is two-fold degenerate and gives two sub-states in a magnetic field. Two basis functions are then needed. They must be spinors and the rotation operators are spinor operators. We may indeed represent the rotation group in 3-space by operations on spinors with two components, and that is the obvious representation to use when a spherically symmetrical atom has two distinct states in a magnetic field. The tangent operators at the origin of the rotation angles are the Pauli matrices, $\sigma_1, \sigma_2, \sigma_3$, and a rotation about three axes through Euler angles α, β and γ is represented by the three matrices

$$\begin{pmatrix} e^{i\alpha} & \cdot \\ \cdot & e^{-i\alpha} \end{pmatrix}, \quad \begin{pmatrix} \cos\frac{1}{2}\beta & \sin\frac{1}{2}\beta \\ -\sin\frac{1}{2}\beta & \cos\frac{1}{2}\beta \end{pmatrix} \text{ and } \begin{pmatrix} \cos\frac{1}{2}\gamma & \sin\frac{1}{2}\gamma \\ \sin\frac{1}{2}\gamma & \cos\frac{1}{2}\gamma \end{pmatrix},$$

operating on a 2-spinor.

If, in conformity with special relativity, we wished to be able to include a boost in this representation we should need to have a fourth matrix with which to multiply the product of the other three in order to obtain the overall operator on a 2-spinor. However it is well known that there are only three 2×2 matrices of the Pauli form so that the required fourth matrix does not exist. If we are to be able to have a representation with both a two-fold axial degeneracy and the possibility of representing a boost, we need to go to a 4-spinor basis. The rotation tangent operators will then have the form $\begin{pmatrix} \sigma_i & \cdot \\ \cdot & \sigma_i \end{pmatrix}$ while the boost tangent operator is $\begin{pmatrix} I & \cdot \\ \cdot & -I \end{pmatrix}$ where I is the 2×2 unit diagonal matrix (see Cook 1988a).

The matrices are 4×4 and so they operate on a 4-spinor as the basis of the representation. That might suggest that the original spherically symmetrical state should split into four distinct states in a magnetic field, but the 4×4 rotation operator is in fact composed of two identical 2×2 matrices on the principal diagonal, so that there are only two distinct eigenenergies (provided we restrict consideration to low energies less than that corresponding to the conversion of the mass of the atom into radiation).

We see that from a consideration of the number of states into which a degenerate state splits in a magnetic field, we may infer the number of basis functions in the representation of the rotation group that corresponds to the physical state. On the face of it a doubly degenerate state requires two distinct basis functions and it is not possible to infer that we have to do with the Lorentz group. Since however we know that a representation of two dimensions cannot be extended to represent the Lorentz group in the way that a representation of dimension 3 can be, we conclude that for consistency we should go to a 4-dimensional representation which we know can represent the Lorentz group, it follows that the equation of motion must have the form of operations on spinors and that the operators must be linear combinations of Dirac 4×4 matrices.

The argument up to this point has been developed for a space of at most four dimensions, one of space and three of time, and that implies that we are restricted to the dynamics of a single particle, for if we are to consider more than one particle we should need additional sets of spatial dimensions to describe their behaviour. To describe the dynamics of two particles we need one time dimension and two sets of three spatial dimensions. I have discussed the formal consequences of that elsewhere (Cook, 1988a). Quantum electrodynamics must also be included (Berestetskii *et al.*, 1982).

4.4 Conclusion

Two arguments have been used to establish a suitable form of kinetic energy for a formulation of quantum mechanics that is Lorentz-covariant, the first based on an assumption about the time of travel of electromagnetic interactions in atoms and the second based on observation and ideas of symmetry and degeneracies. The second approach is much the more fundamental. In the first argument we postulate an internal structure of an atom – nucleus and electrons with electromagnetic interactions – of which we can have no direct knowledge. In the second approach we start from the observations that a single state becomes a number of distinct states when a preferred direction is set up in space. The theory of the representation of groups shows that it is not possible to give a satisfactory account of that phenomenon if we restrict ourselves to three dimensions, irrespective of whatever may be the internal structure of the atom. At that point we may return to the fundamental time-evolution equation of quantum mechanics, in which the time derivative is of first order, and argue that if the kinetic energy is to be a function of the momenta, it must be a linear combination of the momenta. Thus we return to Dirac's form of the equation of motion for a single particle in an external field.

The arguments we have employed in the second approach demonstrate the great power and scope of group theory in physics. The theory of groups and their representations has many specific applications in physics but here we have made use of some quite elementary properties of representations to derive a result of great significance, namely that the geometry of special relativity is the correct one to use in making models of microphysical systems. In Chapter 7 the application of group theory to theoretical physics will be extended in the most general manner.

5

Indeterminacy in theory and observation

5.1 Introduction

I HAVE ARGUED in the preceding chapters that certain key elements of quantum mechanics and special relativity are the consequences of the ways in which we establish standards and make measurements; as such they have clear unambiguous postulates from which rigorous conclusions follow. Classical dynamics and electromagnetic theory are likewise clear and rigorous. Those models may not, probably do not, correspond exactly to the actual physical world, but they are logically sound and consistent and we see clearly their foundations and consequences. Not all theoretical physics is so tidy and some important topics are subject to significant uncertainties which in part are the consequences of inadequate theoretical models, but in part are inherent in the physics itself and the measurements that it is possible to make. The origins of uncertainty in classical physics and the implications for theoretical physics are set out in this chapter. It is important to distinguish between defects in theory and real unpredictability of the physics and to appreciate that if some physical system exists, notwithstanding that an adequate theory has not yet been developed, then the difficulties reside either in deficiencies of the models or in inadequate mathematical treatment. Nature produces a particular physical system but we may not have the insight or technique to generate an adequate model. So long as we are defeated by nature in that way, theoretical physics is not dead; at the same time the behaviour of some systems is truly unpredictable.

Until recently the study of physics was to a large extent restricted to systems that behave linearly and predictably in the sense that their evolution in time is related in a regular way to the starting conditions; it has been very successful in elucidating many major aspects of our observations of the world around us. Not all physics is so straightforward and there are notable subjects where the evolution in time cannot be predicted from the starting point and the dynamical equations. That is the consequence of non-linear interactions or processes and is very common, even though much of physics can be treated as if the behaviour were simply linear.

A consideration of classical dynamics poses the questions we discuss. Classical dynamics is the study of how systems evolve in time and Henri Poincaré (1916) is principally responsible for our present understanding of its essential features. Let a system be specified by a number of independent variables, x_i, which might be canonical coordinates and momenta; they may be thought of as components of a vector $\mathbf{x}$ in a suitable multi-dimensional space. The rates of change of each variable are supposed to be given by functions, X_i, of some or all of the variables; they are components of a vector $\mathbf{X}(\mathbf{x})$ which is a function of the vector $\mathbf{x}$. Thus we have the set of equations

$$\frac{\mathrm{d}x_i}{\mathrm{d}t} = X_i(x_1, x_2 \ldots x_n),$$

or in vector form

$$\frac{\mathrm{d}\mathbf{x}}{\mathrm{d}t} = \mathbf{X}(\mathbf{x}).$$

Such equations generate a dynamic flow. They were first developed for mechanics but now cover not only all classical physics but parts of biology, economics, and other fields as well. We have already seen that somewhat similar equations apply in quantum mechanics; they determine the wavefunction, but then observable quantities have to be derived from operations on the wavefunction and it is at that stage that quantum uncertainties come in. We shall in this chapter deal with classical dynamics.

Three questions, among others, are important: are the definitions of variables and of functions unique, how do the solutions depend on the initial conditions, do the equations have unique solutions?

We saw in Chapter 3 that coordinates and functions of them, such as potentials, by which we describe the state of a distant physical system, are not unique; nonetheless they cannot be changed in arbitrary ways. Different sets of coordinates and functions must be related in such a way that the quantities that we can measure, namely times of emission and reception of electromagnetic signals and so on, remain invariant. The proper relation is of course the Lorentz transformation. There are also circumstances, especially in biology and economics, where it may not be obvious what variables should be chosen to specify the state of a system. In this chapter we consider the other two problems.

We commonly expect the solutions of the equations of dynamics to vary continuously with the initial conditions, but as Poincaré (1910) pointed out, that it is not always so and even though definite solutions to the equations can be found, the solutions are so sensitive to the initial conditions that the outcome is essentially unpredictable. The dynamics are then said to be chaotic. There may also be circumstances in which formal solutions to the equations cannot be found. I have distinguished between chaotic and formally insoluble dynamics, but in reality they are often connected.

Chaotic behaviour is such that while the equations of motion of a system admit of exact solutions, either those solutions depend so critically upon initial conditions that in practice the initial conditions cannot be prescribed accurately enough for meaningful results to be obtained, or they are effectively independent of the initial conditions. That is not just a matter of analytical or numerical calculation, the fact that physical behaviour itself depends so sensitively upon conditions has observable physical consequences as well as consequences for the matching of theory and observation.

Henri Poincaré examined the question of whether the solar system was stable over long times and also set out the idea of chaos particularly clearly in discussing meteorology. Chaotic behaviour has since been identified in a wide range of physical phenomena, in biology and in economic systems. Chaotic dynamics are closely related to the statistical behaviour of systems of many bodies and both entail uncertain behaviour of the systems themselves as well as of models of them. The first chapters of this book showed how the certainties of measurement determine the form and scope of theory, the aim of this chapter is to show how the uncertainties of nature determine the form and scope of theory.

5.2 Chaos

In many circumstances a variation of the initial conditions for the solutions of dynamical equations leads to variations of comparable magnitude in the solutions and then it can be said that the results are predictable from the initial conditions by the equations of motion. In other cases however, a small variation in the initial conditions may produce such a large change in the solution that the result cannot be said to be predictable (Poincaré, 1910). It might be that a variation in the initial conditions that would lead to an observable change in the result was so small as to be unmeasurable. It might also be that such a variation would be less than natural fluctuations in the initial conditions. Classical chaos is the compendious term for behaviour which is rigorously according to the laws of classical dynamics but of which the outcome is nonetheless uncertain because it depends so sensitively upon initial or other conditions that it is not possible to determine those conditions with sufficient precision to predict the outcome. For all practical purposes the outcome is independent of the initial conditions or, from another point of view, its relation to them is random.

The mathematical equations that constitute almost all of theoretical physics as it has been developed up to now, are linear, that is to say, they are linear combinations of functions of variables or their differentials or integrals. Linear theories do in fact represent a very great deal of physics very well, the most notable exception being fluid dynamics, for which the equations of motion, the Navier–Stokes equations, are inherently non-linear. The solutions to linear differential or integral equations are linear functions of the initial conditions and so with few exceptions, linear theories are predictive in the sense that Poincaré explained – a change in the initial conditions will give a proportionate change in the solution and the factor of proportionality will not depend on the initial conditions. One important consequence is that linear perturbation theory can be used to calculate differences between solutions for conditions that do not differ greatly.

Linear models of physics can very generally be put in terms of groups of operators operating on vectors over a suitable field and transforming them into other vectors over the same field. The most general group is the general linear (GL) group. Groups by definition, include a unit operator that changes nothing, and every operator has an inverse in the group. The evolution equations that we now consider correspond to operators, often non-linear, that do not have inverses and so form semi-groups, not groups. The theory of semi-groups, and especially that of non-linear semi-groups, is less developed than the theory of groups and their representations, so that it is more difficult to make general arguments, such as are advanced in Chapter 6, based on properties of group representations, about the relation of theory to observation.

A very simple equation that occurs frequently in text books as an example of chaotic behaviour is the continuous logistic differential equation,

$$\frac{\mathrm{d}x}{\mathrm{d}t} = ax(1-bx);$$

its solutions show a number of the characteristic features of chaotic behaviour. Initially the value of x increases exponentially and is proportional to the initial value, x_0, but at large times $x(t)$ tends to $1/b$, independently of the value of x_0. The discrete logistic difference equation shows the further complication that for small values of a there is just one solution, but as a increases first two and then successively more solutions become possible, between which the value of $x(t)$ switches, until the behaviour appears completely random. The example is instructive in three ways, it shows that the relation of the solution to the initial conditions depends on the initial conditions, it shows that in certain ranges of the independent variable (t) and for certain values of the initial conditions, the solution is independent of the initial conditions, and finally, it shows that the type of solution depends on the values of parameters in the equation. Many non-linear equations display those characteristics.

Equations in two or more variables introduce a further feature. Suppose the equations for two variables are

$$\frac{dx}{dt} = ax - bxy,$$

$$\frac{dy}{dt} = cy - dxy.$$

Solutions for which dx/dt and dy/dt are zero are $\{x = 0, y = 0\}$ and $\{x = c/d, y = a/b\}$; such solutions are known as fixed-point solutions.

Near the fixed point $\{0, 0\}$, x is proportional to $\exp(at)$ and y is proportional to $\exp(ct)$ and if either a or c is positive, the solution is unstable.

If (ξ, η) are small displacements from the solution $\{c/d, a/b\}$, they are of the form

$$(\xi, \eta) = (\xi_0, \eta_0) \exp[\pm(ac)^{\frac{1}{2}}t].$$

If a and c are of the same sign, the product ac is greater than 0, the square root is real and one term in the solution increases with time so that the solution is unstable. If, on the other hand, a and c are of opposite sign, the solutions are oscillatory, which means that ξ and η map out a trajectory that circulates around the fixed point.

Those equations display a further property of non-linear equations – as a parameter changes continuously, the character of the solution may undergo a discontinuous change at some value of the parameter. Here if one or other of a or c passes continuously through zero from one sign to the other, the solution will change from being stable to unstable or vice versa.

The form of a linear equation is independent of the boundary conditions and of the region of the space of variables in which the solution lies. The form of a non-linear equation, or at least the values of the parameters in it, will in general vary with the initial conditions and with the range of the solution.

Chaotic behaviour occurs as a result of an interaction between the local and global conditions on the dynamical equations. Consider an equation of the form

$$\frac{\mathrm{d}x}{\mathrm{d}t} = ax + b,$$

in which a and b are functions of the variables x, y.

When t is small, x behaves as $x_0 \exp(at)$, where x_0 is the initial value of x and a is evaluated with x_0 and the initial value of y. We see that a difference δx_0 between neighbouring initial values of x_0 grows exponentially with time. That, however, is not sufficient for the behaviour to be chaotic. If a were constant, the effect of a variation in the initial conditions would be known at any subsequent time, even though the deviation increases exponentially with time. Chaos requires a further condition, that the range of the variables, $x, y\ldots$, should be bounded so that the exponential solution cannot continue indefinitely.

In reality all physical systems are bounded in some way, although the bounds may not always be very stringent. Restraints on the variables themselves, the energy available to drive a system, are examples of the ways in which systems are bounded. Biological systems also are bounded, by the space accessible to a population or by the nutrient supply, for example. Sometimes the dynamical equations themselves lead to appropriate bounds, but they may not always do so and some additional global condition, which can often be imposed by a modification to the dynamical equations, applies to the solutions.

It is helpful in further discussions to use geometrical representations. Suppose there are n variables, x_i. They are all functions of time, and so a solution to the dynamical equations traces out a one-parameter curve $\mathbf{x}(t)$ in the phase space of the variables. We call that the trajectory of the solution and can think of the rates of change of the variables on it as defining a tangent to it, with a direction corresponding to increasing time.

Limits on the values of the variables lead to solutions that are independent of initial conditions. Suppose that the rate of change of a particular variable, x_i, vanishes on some manifold $\{X_i = 0\}$ in the space of the variables (Figure 5.1) (with two variables the manifold would be a line, with three variables a surface). Let that

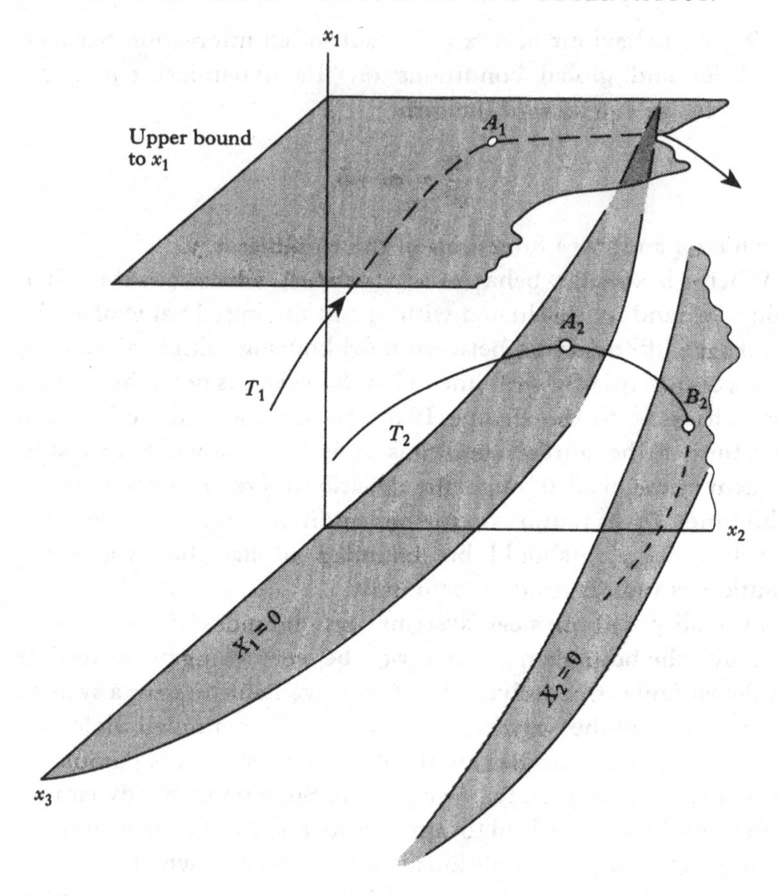

Trajectory T_1: meets upper bound on x_1 at A_1;
T_2: $\dot{x}_1 = 0$ at A_2,
$\dot{x}_2 = 0$ at B_2

Figure 5.1. Trajectories of solutions of differential equations in phase space of three variables, with upper bound to variable x_1 and manifolds on which X_1 and X_2 vanish.

manifold intersect a boundary which sets a limit to x_i, and on which therefore X_i vanishes and within which X_i is negative. The intersection will be in a point if there are two variables, on a line with three variables and so on. If a solution never attains the bounds it may, as just indicated, vary continuously with the initial conditions in a predictable manner. If however it attains the bound, that means that the representative point of the solution will remain stably on the boundary until it reaches the manifold $\{X_i = 0\}$. Thereupon the sign of X_i changes, the solution becomes unstable with respect to deviations towards the interior of the boundary and the solution point will move into the interior of the domain of the solution. In two dimensions it will start from a fixed point with X_i zero and the other rate of change determined by the values of the variables at the intersection. The trajectory of the solution thereafter is independent of the initial conditions before the boundary. With more than two variables there will be a range of points at which the trajectory of the solution can leave the intersection, entailing some dependence on the initial conditions before the boundary. Trajectories after the intersection are determined by conditions on the intersection; with two variables there is just one possible trajectory but with more variables there is a number of possible trajectories.

These last observations indicate that the behaviour of a non-linear system is not necessarily completely random, for a boundary imposes a limit to the possible trajectories of solutions after leaving a boundary. We go on to look more generally at factors that determine common properties of the solutions of some non-linear systems.

The fact that the logistic equation has solutions that are in certain circumstances independent of the initial conditions, and not just very sensitively dependent on them, means that the behaviour is controlled by internal parameters of the physics such as the coefficient b in the logistic equation. In other instances it seems that external constraints, such as limiting values but not initial conditions, control the behaviour. The classical perfect gas is an example. The internal motions of the molecules are wholly indeterminate, it would seem, yet they must be such that the total momentum and angular momentum are zero in the absence of external forces, while the internal energy must correspond to the ambient temperature. How are the velocities of the individual molecules adjusted so that those conditions are satisfied?

Fluid dynamics afford further examples. Harold Jeffreys pointed out many years ago (Jeffreys, 1933) that the steady winds over the surface of the Earth are such that on account of friction at the surface, torques are exerted that transport angular momentum across latitudes. He argued that there must be some compensating transport of angular momentum in the reverse direction because there were no torques acting on the solid Earth and the atmosphere to give rise to the transport in the winds. He showed that the required compensation was provided by cyclones carrying angular momentum from low to high latitudes. Now cyclones are chaotic phenomena, their origin is erratic and so are their subsequent paths, yet overall, the total transport of angular momentum must balance that by the steady winds. How is it achieved? Similar questions arise about eddies in the oceans. Presumably if there is transport of angular momentum across latitudes, there would be a corresponding change of angular velocity in the system, as with gyroscopic dynamics, which would stimulate the generation of cyclones. In fact the balance is not quite exact and there are small fluctuations in the rate of rotation of the Earth corresponding to net changes of angular momentum of the atmosphere (and presumably of the seas).

Similar arguments apply to other chaotic systems – general dynamical principles ensure that certain integrals of the motion are either constants or change in a way determined by external constraints. The essential principles of dynamics are not violated in chaotic systems, and that means that there should be meaningful quantities to be measured and meaningful physics to be done with them.

The nature of chaotic dynamics itself also seems to ensure that there are meaningful measurable quantities. Chaotic systems are bound systems, the constituents of the system are not independent, and so it should be possible to identify some properties of the system as a whole from which the values of the individual variables deviate. The system properties will be those that can be measured by external observers. A perfect gas affords an example. The total momentum, angular momentum and energy are well defined by observation and can be measured, but the deviations of the positions and velocities of the molecules from their mean values are inaccessible to observation. A bound system is also a bounded system, that is, the ranges of the variables are limited in some way, which means that there will be functions of the variables of which the integrals over the range of variables are also bounded. Such integrals will not be constants, but their averages over a range of initial conditions may be found to be stable.

5.3 Observables and predictions of chaotic dynamics

The principle of special relativity (Chapter 3) follows from the requirement that functions of observations should be invariant under transformations that are applied to variables that specify the state of a system – for example, the space-time interval and the rest energy must not change. A similar principle of invariance, or relativity, or conservation, known as Noether's theorem, applies whenever the dynamical equations can be derived from an action principle. There is then a Lagrangian, L, of the system and the equations of motion, of time evolution, are

$$\frac{\mathrm{d}}{\mathrm{d}t}\left(\frac{\partial L}{\partial \mathbf{v}}\right)-\frac{\partial L}{\partial \mathbf{x}}=0,$$

where $\mathbf{x}$ is a vector of variables and $\mathbf{v}$ is the vector of their rates of change, $\partial \mathbf{x}/\partial t$.

Let us put $\mathbf{x}$ equal to an arbitrary constant ξ plus deviations $\mathbf{x}'$, where ξ is to be considered as the arbitrary origin of coordinates. We suppose that the behaviour of the system is independent of ξ. Put $\mathbf{u}$ equal to $\partial\xi/\partial t$, so that $\partial L/\partial \mathbf{u}$ is the momentum $\mathbf{p}_\xi$ corresponding to the vector ξ. Because L is independent of ξ, $\partial L/\partial\xi$ is zero and so by the Lagrangian equations, the momentum $\mathbf{p}_\xi$ is constant in time.

Commonly the Lagrangian is taken to be the difference of kinetic and potential energies, $(T-V)$, where both T and V are smooth functions of the variables, but there are other possibilities. For instance, the Lagrangian of a perfect gas in an enclosure might be

$$\sum_i mv_i^2+\int_T V\mathrm{d}\tau,$$

where T is the volume of the enclosure, V the potential energy density and $\mathrm{d}\tau$ the element of volume.

If V is zero inside the enclosure but is very large (V_S) at the boundary, S, then the Lagrangian is

$$\sum_i mv_i^2+\int_S V_S\cdot\mathrm{d}S,$$

an expression that is independent of the origin of coordinates and time so that the total energy and momentum of the gas are conserved (the total momentum of a gas in a stationary enclosure is of course zero).

Energy and momentum are conserved in any system when the observations are independent of the origins of time and position. Other integrals of the motion will be conserved if a Lagrangian is independent of the origins of the corresponding variables. That may be so for the overall behaviour of a chaotic system and the integrals of such a system that correspond to the total energy, momentum and angular momentum and other quantities must then be constants. However not all systems can be described by variables with arbitrary origins – densities and populations, for example, have a unique origin at zero and in such cases there may be no conserved momentum nor energy.

The aim of this book is to tease out relations of theory to observation – what can be observed and how does that constrain theory? Formally the answer to that question for chaotic systems seems to lie in the remark that chaotic systems are not truly autonomous, although as seen above, they may at first appear to be. They are subject to external constraints and to laws of conservation – the total energy may be conserved, total momentum and angular momentum may be conserved, as may be numbers of particles and mass. Because such quantities have definite conserved values, they should be measurable and an empirical theoretical physics of a chaotic system should be based on them – the classical example is the axiomatic theory of thermodynamics which reproduces the external macroscopic behaviour of systems without detailed knowledge of the internal microscopic chaotic dynamics.

It may not always be so clear as it is with mechanical systems just what quantities are conserved, nor the means by which conservation is effected, and one aim of the study of chaotic systems should be to identify properties of a system or functionals of the mathematics that should be conserved and therefore measurable.

Prediction, in Poincaré's sense of the connection between the solution of a set of differential equations and the initial conditions, is not possible in chaotic systems. That is the definition of such systems – the end point of a dynamical flow is, so far as meaningful observation goes, unrelated to the starting point of the flow. That does not mean that no prediction is possible about the evolution in time of a chaotic system. In the first place, the solutions of a set of equations may depend only on the parameters of the equations and not at all on any conditions. The continuous logistic equation is an example – after a sufficient time the solution as a function of time, depends only on the parameters of the equation. Other chaotic time series behave in ways that are not purely random but are independent of the starting conditions.

Systems of equations for more than one variable may have strange attractors and basins of attraction, with solutions that remain within a certain range of values of the variables and which, after a sufficient time, will be independent of initial conditions. There will be predictable features of the motion, such as the ranges of values, the energy of the motion and so on.

A third form of predictability occurs in ordinary statistical mechanics. When the numbers of elements of a system are very large, the probability of the most probable value of a mean value such as the energy is so great as to amount almost to certainty/(Waldram, 1985). It is then possible to construct an empirical physics of the relations between such mean values. Further, since quantities such as the total energy are determined by external conditions, as already explained, relations between them become predictable in the sense of Poincaré.

Chaotic systems are unpredictable but that does not mean that they are wholly erratic. They may be subject to external constraints, they may be autonomous but controlled by internal parameters. A physics of chaotic systems will have an aim of identifying those aspects of the behaviour which are in some sense predictable.

5.4 Systems of many bodies

A second reason for uncertainty in physics is that it is not possible to obtain definite solutions to the dynamical equations, and we consider circumstances in which the dynamical equations may turn out to be insoluble in the dynamics of systems of many interacting bodies. No real physical system is just two bodies. One-electron atoms are never isolated in the real world but are always subject to the fields of other atoms or molecules. Tangible matter in any form is made up of very many particles which are not independent but bear some relation to each other and form a recognisable entity. The physics of systems of many bodies is therefore a very large part of the study of the world around us, and so we have to confront the particular difficulties of applying general dynamics to systems of many bodies. We may face the problem of three bodies attracting each other gravitationally, a problem of which the motion of the Moon about the Earth and Sun is the classical instance and still one of the most difficult and demanding. We may be concerned with the structure of metals or the properties of plasmas; we may wish to study the behaviour of liquids or the statistical mechanics of condensed matter generally.

The solar system exists, metals exist, liquids exist, we are able to make unambiguous measurements of their macroscopic properties and there is a well developed physics of metals and liquids that relates observed phenomena empirically without necessarily making use of a detailed theoretical model of the internal microphysics. Classical thermodynamics likewise is very effective on a macroscopic scale even if the details of the dynamics on the microscopic scale are beyond us. There are thus properties of such systems that can be related to each other by an appropriate macroscopic theory with its own premisses, even though they cannot be calculated by an *a priori* microscopic theory on a more general basis common to other parts of physics. We shall consider how far predictability in the gross can be restored to systems that are in detail unpredictable and correspondingly how far there are properties of such systems of which meaningful observations can be made.

The root of the difficulties that arise in the theories of all those systems and states of matter is that it is not possible to obtain explicit solutions of the equations of motion, whether in classical or in quantum mechanics. The classical problem, which arose first in celestial mechanics, is to determine the motions of a set of bodies (n in number) attracting each other by the force of gravity. If there exists an Hamiltonian there will be $6n$ first order differential equations to solve for the $3n$ coordinates and $3n$ momenta as functions of the time. If a definite solution is to be found, then $6n$ initial conditions will have to be prescribed. In celestial mechanics it might be thought that should be possible, for the positions and velocities of all the bodies can be determined at some initial instant and the equations of motion might then be integrated numerically to find later positions and velocities. In fact that is done in the practical numerical calculation of the motions of the heavenly bodies, of space craft and of artificial satellites, but the solutions are reliable only for rather limited times because numerical calculations are not free of errors which build up after many iterations. The calculator needs some control over her results after the lapse of time, she would like to be able to check her results against a formal analytical solution which would apply to any initial conditions and not just the particular ones of the specific problem. Particular solutions can often be made numerically, general solutions are impossible for three bodies or more.

When the three components of the initial position of the centre of mass are added to the quantities that are constants in the absence of external forces, namely the total energy, the three components of angular momentum and the three components of linear momentum, together with two other arbitrary constants it is found that there are 12 constants of integration in all (Whittaker, 1927). Now the total number of unknowns in a system of two bodies is also 12 and consequently the motions of the two bodies are determined for all time by the constants (or integrals) of the motion. The simplest example is the Keplerian elliptical orbit of two bodies moving around each other under their mutual gravitational attraction – the orbit is an ellipse of constant size and eccentricity and with a fixed orientation in space, described at a constant rate. Those geometrical constants are determined by the dynamical constants.

As soon as just one additional body is included, the problem becomes indeterminate – at any instant there is a whole range of possible positions and motions consistent with the initial positions and velocities and the dynamical integrals. The classical problem of three bodies is strictly speaking insoluble and that is why the dynamical problem of the Moon's motion is so very difficult (the only one, said Newton, that made his head ache). It is also why it is not possible to prove that the solar system is stable over a long time. If the positions of three or more bodies are to be found from those at an earlier instant, it is necessary to know the accelerations and therefore the forces between them, but the forces between the bodies cannot be calculated without knowing the positions, which are still to be calculated. The corresponding problem for two bodies can be solved because the two positions of the bodies reduce to one distance between them, together with the motion of the centre of mass, which is independent of the mutual attraction.

The argument from the number of integrals of the equations of motion does not involve the linearity or non-linearity of the equations, but it may easily be understood that equations for many bodies are often non-linear. In celestial mechanics, for instance, the distance between two bodies is a non-linear function of their own coordinates, which themselves may be functions of the coordinates of other bodies; the relative acceleration is consequently in general a non-linear function of the coordinates of a number of bodies. Almost all quantum systems are many-body systems and for that reason are formally insoluble just as are classical systems, although quantum systems do not exhibit chaos as classical systems do.

Some very successful theories of systems of many bodies exist, but at the same time there are notable difficulties. It cannot be said that a fully satisfactory theory of the liquid state exists, and the relativistic dynamics of systems of many bodies, both quantum and classical, are not fully understood. There is one important difference between the classical problems of celestial mechanics with just a few bodies, and problems involving very many bodies, whether in classical or quantum mechanics. In the former it is possible to observe the positions and velocities of all constituents (for example, every planet and satellite of the solar system) at any time and the theoretical requirement is to calculate them, even though classical theory tells us that is not possible analytically. On the other hand the positions and velocities of every particle of a classical plasma or perfect gas cannot be observed, nor can they be in an atom nor in a molecule; the theoretical requirement now is to find the gross macroscopic consequences of states that cannot be observed in detail.

Although the problem of three or more bodies has no solution in general, yet it is possible to obtain results in the theory of three and more bodies that correspond to observation, though only by restricting the problem. The next section indicates some of the ways in which that has been done.

Traditionally the solution to a problem of a few bodies has been obtained through some perturbation and much of the history of classical dynamics is the development of principles of perturbation methods in lunar and planetary theory, together with their applications. Those matters will not be considered further here, for they go far beyond the range of this book (but see Cook 1988b); enough to say that perturbation methods have the same limitation as numerical methods, their stability over long periods of time is not always assured.

5.5 Nearly soluble problems of many bodies

There are essentially two ways to reduce a problem of many bodies to make it soluble. One is to impose some geometrical boundary condition which enables the positions and velocities to be known independently of the solution of the equations of motion. The other approach is to suppose that the influence of some of the constituent bodies on the others is very small – in atomic and molecular physics that is the Born–Oppenheimer principle; the first examples of both approaches occurred in celestial mechanics.

In the solar system there are instances of small bodies that maintain the same geometrical relation to two much more massive bodies which themselves move in a mutual Keplerian orbit. The positions of the small bodies are called the Lagrangian points (Figure 5.2), and they are stable, for once a small body gets to such a point, frictional dissipation generally reduces its velocity so that it remains there. Here there is a geometrical constraint upon the system – if the small body is at a Lagrangian point, its position and velocity are known. However, that can only be within some limit upon the size of the small body relative to the other two, for it must not disturb their mutual orbits.

Lunar theory provides the other example. Here there is a single very massive body, the Sun, at a very great distance, and the Earth and the Moon execute orbits about each other in its gravitational field. The simplified problem is soluble because the attraction of the Sun can be considered as known independently of the motions of the Earth and the Moon – even so the theory is difficult and the observations of the Moon are so precise that allowance has to be made for the fact that the Sun is not infinitely massive and distant,

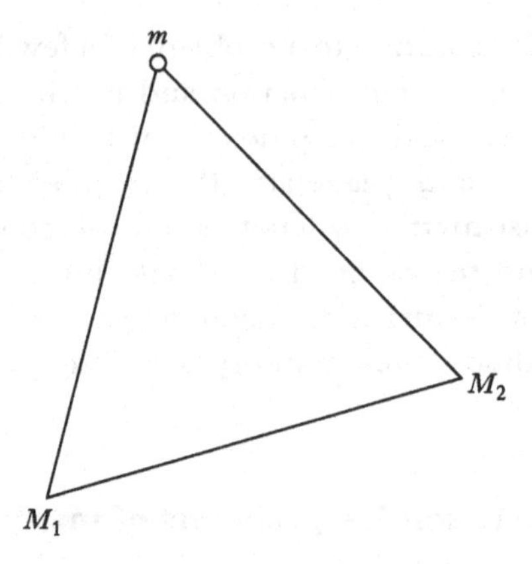

The small body, mass m, is at the vertex of the equilateral triangle $M_1 M_2 m$.

Figure 5.2. Stable Lagrangian triangle of small body, mass m, in the gravitational field of two massive bodies, M_1 and M_2.

as well as for many other deviations from the basic model (see Cook, 1988b).

If the particles at a Lagrangian point are very small, many may occupy the neighbourhood of the point because their mutual interactions are so weak that they do not affect the dynamics – the small particles behave independently of each other. Similarly, small artificial satellites of the Earth follow the lunar theory independently because again, their mutual interactions are not strong enough to affect each others' behaviour. A further way of reducing the problem of many bodies to a manageable form is therefore to find if possible some bodies or groups of bodies that behave independently of others. That cannot be taken too far because a system of many bodies has no significance unless it exists as some recognisable entity, as a bound system, over an appreciable period of time – bodies that are truly independent of each other do not constitute bound systems. The small bodies of the solar system are bound systems because they are in a common potential, not because they interact directly with each other; but they form rather special systems.

The normal modes of a vibrating system are, to first order, independent and are assumed to be so in the theory. Normal modes occur in systems governed by the Hamiltonian equations of motion, having no dissipation and with the kinetic energy proportional to the squares of the velocities and the potential energy proportional to the squares of the displacements from positions of equilibrium. Although these conditions may appear restrictive, many classical and quantum systems satisfy them quite closely, so that normal modes are a good representation of the dynamical behaviour of whirling shafts, suspensions of motor cars, the free elastic oscillations of the Earth as a whole, the vibrations of molecules and phonons in crystals, among many others.

An essential property of normal modes is that they are independent. Once the masses, the force constants (or their equivalents in a continuum) and geometrical constraints are known, the forms of the eigenfunctions and the values of the characteristic frequencies can be calculated, while the relative amplitudes are determined by the initial conditions of velocity or acceleration. Thereafter each mode behaves independently. The position and velocity of each particle in the system can then be found by superposing the amplitudes of the normal modes at the position of the particle. Strictly speaking the positions and velocities of individual particles are as much constructs of the theoretical imagination as are coordinates of distant objects in special relativity, the observable quantities are amplitudes and frequencies of Fourier components of the oscillations.

In reality normal modes are not truly independent. Coupling can often be ignored and then the independent normal mode model is an adequate representation of the physics but there are also situations in which it cannot be ignored. The thermal energy of a solid is the energy of the elastic vibrations, the phonons. If they were truly independent, their relative amplitudes would be determined by some arbitrary external conditions. Instead, in most ordinary circumstances, there is thermal equilibrium with the amplitudes following the Boltzmann distribution. It is coupling between the normal modes that brings about that state of affairs and the relaxation time to thermal equilibrium is commonly very short.

The structure and dynamics of a molecular crystal is a problem of many bodies which separates into two parts, the static equilibrium structure and the elastic vibrations. The static structure can be determined through an appeal to symmetry because the relative positions of all particles are known from the space group of the crystal and the way it repeats throughout the crystal. If the law of force is also given, the energy can then vary only with the absolute dimension of the unit cell. The vibrations may be resolved into normal modes as just discussed, and again symmetry determines the possible modes out of all those that can exist in an otherwise structureless solid. The structure and vibrations of the ionic framework of a molecule are governed by similar considerations of symmetry, except that now the symmetry is that of point groups instead of space groups. Appeal to symmetry is a very powerful way of avoiding the ambiguities of many-body theory because symmetry fixes the relative positions of constituents and so reduces the number of independent coordinates from $3n$ to 3.

In the soluble three-body problems of celestial mechanics one at least of the bodies is small enough that it does not influence the others. We may then calculate the motions of the larger bodies independently of the smaller one and so find the potential within which the small body moves. The Born–Oppenheimer principle applies the same idea in microscopic physics. In its original form it stated that the ions in a molecule could be considered to be stationary in working out the motions of the electrons because the ions are so much heavier than the electrons and therefore move so much more slowly. Since the positions of the ions conform to the appropriate point symmetry group, there is just one scale parameter unspecified in the potential in which the electrons move. The ions themselves move in a potential which is built up of their mutual repulsion and the attraction of the electrons. The former again is determined by the point group symmetry and the scale factor, while the latter is regarded as an average over many characteristic periods of the electronic motions (which themselves must conform to the same point group symmetry). In that way the dynamics of the ions and electrons are separated into calculable parts with an interaction that is governed by symmetry and a single scale factor. The Born–Oppenheimer principle also governs the dynamics of a metallic solid which, like a molecule, has an ionic framework within the potential of which there is a cloud of electrons. The only differences are that the symmetry is that of a space group instead of a point group and that the Born–Oppenheimer conditions are more fully satisfied because of the relatively greater mass of a crystal.

When the ions in molecules are treated as stationary it is necessary to know their individual positions in order to calculate the potential in which the electrons move. The ions, on the other hand, are in some average potential of the electrons and the individual positions of the electrons (whatever that may mean) are not required. The observable properties of the electronic cloud, its energy and angular momentum, and the potentials at the ions, are average properties over all the electrons, and can be calculated as matrix elements with respect to the wavefunction of all the electrons collectively without knowing the instantaneous positions of the electrons separately.

The potential to which the ions in a metal are subject is some form of average over all the electrons. The electrons can be treated collectively by classical electrodynamics as a continuous polarizable medium having a complex susceptibility; they can be described by a single field instead of as a collection of interacting particles. The properties of the medium follow from the fact that electrons are fermions and so must satisfy the exclusion principle. The ionic charges produce an external field to which the electronic medium is subject and self-consistent methods of classical electrodynamics are used to obtain the response of the electrons (Lindhard, 1954). Thus the energy of the electrons in the field of the lattice of ions is calculated. A theory that concentrates on the overall cooperative properties of the electrons treated as a plasma corresponds very well to the observational circumstances that plasma properties of a metal can be observed (or rather, that certain observations can be interpreted as those of plasma properties) while the motions of individual electrons cannot be distinguished. Superconductivity is a special case in which one particular plasma mode of the electrons dominates all others.

Electrons in metals bring us to the boundaries of statistical mechanics and classical thermodynamics. No attempt is made to calculate the positions or velocities of separate electrons, the only quantities that are found are collective properties of the electrons such as the polarisability or susceptibility, and the energy. In the first approximation, at least, the electrons are supposed to have no thermal energy, but most substances in most circumstances are not at zero temperature and have finite thermal energy. Thermodynamics is the study of those collective properties of systems which depend crucially upon temperature in circumstances such that the behaviour of the elementary constituents can only be found statistically.

We normally associate statistical mechanics with very large numbers of particles or constituents or possible systems. The need to treat the system statistically arises because the number of components is large. Non-linear dynamics leads to statistical treatment in a somewhat different way. When the number of variables in a non-linear system is small, there is a unique trajectory for the solution of the dynamical equations, even though the relation of that single trajectory to the initial conditions may be indefinite. When however, the number of variables is large it turns out that there are alternative solution trajectories for a single set of initial conditions, even discounting the effect of boundaries. It appears likely that if the number of variables is n, the number of trajectories is of order $(n+1)!/(2n+1)$, which is about 1 for $n = 3$ but is just over 7000 for $n = 10$. For that reason alone, even small non-linear systems may require a statistical treatment.

5.6 Thermodynamics and statistical mechanics

We have seen that it is not possible to calculate analytically in detail the dynamics of a system of a very large number of constituents; it is impossible in theory and unrealistic in practice. The studies of classical chaos in recent years have given emphasis to the point. Although the equations of motion of a classical system may be well defined, it may be unrealistic to think of measuring the initial conditions with sufficient precision that the solution would be determinate. It is then only possible to state both the initial conditions and the results in terms of probabilities. Large systems do however have collective properties that are sums or averages or other functionals of the dynamical properties of all the individual constituents. The energy, the pressure exerted on a container, the magnetisation and the temperature, just like the polarisability of electrons in a metal, are all collective properties that are well defined statistically. By experiment and theory we can establish physical connexions between them without knowing the details of the underlying dynamics, so that they have an autonomous physics. Much of observable physics indeed is to do with such collective properties and not with internal details, and it is important to develop theories in which those properties are treated as objects in their own right.

Collective properties do nonetheless depend upon the internal dynamics of the system and one of them, the energy, is an integral of the equations of motion for a conservative system. If we suppose that the system is at rest, the other integrals of total momentum and angular momentum will be zero, and thus the energy is the only non-zero integral of the system. Other properties such as pressure depend upon the details of the internal dynamics and since those are unknown, it might seem that it is not possible to calculate a pressure. However, we can in principle calculate the pressure for all possible configurations of the system consistent with the energy, make some estimate of which configuration is the most probable and take the corresponding pressure to be the one that is the observed outcome of the operations that we call measuring pressure. We deal with probable values of collective quantities which are in practice very well defined, both theoretically and operationally by observation, on account of the high probability of the most probable mean values. We do not go behind the results of operationally defined observations.

We have to deal with the statistics of very large numbers. What is the most probable state of a system of very many constituents and how closely is it approached? It is shewn in textbooks of statistical mechanics (for instance Waldram, 1985) that the chance of a deviation from the most probable state is very small, of the order of $1/N^{\frac{1}{2}}$ for a system of N components, and consequently we may take the most probable value to be the one that is found in practice. While that statement seems almost intuitive, it depends on the assertion, the ergodic theorem, that a system will, given long enough, attain every state that is dynamically open to it. The theorem, while plausible, has never been rigorously established, and the usual treatment of statistical thermodynamics proceeds on the basis that the ergodic hypothesis applies in almost all cases. Thus at the heart of statistical mechanics, there is an assumption rather than a demonstration.

The justification for the ergodic hypothesis depends on two propositions. The first is that the states accessible to a dynamical system are distributed continuously in the neighbourhood of any one of them. That is just a statement about the continuity of nature. The second proposition is that any state can be reached by a dynamical trajectory from any other. A dynamical trajectory is one that is according to the principle of least action. That assertion, it seems, has never been proved rigorously; what has been argued is that given any two states, a trajectory starting from one can be found that will pass through a state as close as desired to the second. The difficulty lies in proving that it is always possible to find a trajectory which starts from a given state and leads to a second given state, when the trajectory must depend on the conditions of the first state.

The argument is different for a chaotic system. As before, we may suppose that all possible states are accessible to the system. Now let the system be chaotic in the sense that the final state of a trajectory varies very greatly with the initial state. Suppose in fact, that the limits within which an initial state would have to be specified for a particular final state to be attained, are exceeded by the fluctuations between states of the system that result from chaotic dynamics. That means that states cannot be regarded as points in a Hilbert space but rather that they extend round points of the space. It also means that the end point of a trajectory can be considered to be independent of the initial point since the latter cannot be specified. Thus we do not have to consider whether a particular state can be reached dynamically from another, but only whether it is an allowable state of the system. Since probabilities are evaluated only over allowable states, it can be seen that the ergodic hypothesis is established for chaotic systems.

If we are to calculate probabilities for some model, we must evidently know what states of a system of given energy are accessible to it according to the model. The particles of a classical perfect gas may take any set of velocities subject only to the restriction that the total kinetic energy should be equal to the prescribed energy of the system. Otherwise there is no restriction. The probability of a particle having a specified momentum is given by Maxwell's distribution and any number of particles may have the same momentum, subject only to the restriction on the total energy. Counting numbers of states then reduces to integrations over ranges of velocity. Quantum mechanical systems are different. We know that the total wavefunction of a system must be either symmetrical (for bosons with integral spin) or antisymmetrical (for fermions with half-integral spin) and so there are limits on what states are accessible, as there are not in a classical system. In particular no two fermions can be in the same state. The corresponding distributions are the Bose–Einstein distribution for bosons and the Fermi–Dirac distribution for fermions, and they differ from each other and from the Maxwell distribution especially at very low temperatures.

5.7 Observables of many-body dynamics

Some of the relations between the gross properties of a system are dynamical, thus the pressure of a gas is the derivative of the energy with respect to the volume, $\partial E/\partial V$. Others involve the temperature. Others involve the probability of obtaining a particular state, so that some measure of probability is needed. It is provided by the entropy. Probabilities are ordinarily combined by multiplication but it is convenient to have a measure of probability that, like energy or volume or magnetisation, is additive with the number of constituents of a system. Since the resultant of a number of independent probabilities is their product, the entropy is taken to be proportional to the logarithm of a probability.

While the probability of obtaining the actual value of some system is exceedingly high, the number of states in which a system may exist is also very great and there will be a finite probability of the system being in a state slightly different from that of maximum probability. The entropy measures the range of states around the one of maximum probability within which the system may fall. If the most probable state is certain, the range is zero and so is the entropy; the entropy increases with the uncertainty of the most probable state.

Observations on systems with properties defined statistically seem especially open to the critique that they are heavily indebted to theory. We assert, on theoretical grounds, that the force we observe (with a manometer for instance) on some area of a vessel containing a gas, is the integral of all the changes of momentum of the molecules impinging on that surface. We may likewise set out the theoretical elements in measurements of viscosity, of thermal conductivity or electromagnetic properties. Those measurements usually have stable well-defined values which are explained, again on the basis of theory, as a consequence of the very large numbers of events that contribute to the gross observation and the gross properties of a system are defined by the operations that are made on that system to measure them. The properties are overall statistical properties because the methods of measurement deal with the whole system.

5.8 Summary and conclusion

The dynamics of systems of many bodies are formally insoluble in general, but it is possible to make simplifications that lead to practicable calculations. It may be possible to find combinations of coordinates and momenta that behave almost independently, such as normal modes of vibration; and it may be possible to call upon symmetry to reduce the number of variables to a manageable number. The observable properties of such systems are collective, such as the amplitude and frequency of a particular mode of a mechanical structure or a molecular spectrum. It is natural that theory should follow observation and deal with collective mathematical objects, the eigenfunctions and eigenvalues of normal modes.

The decomposition into normal modes is never exact. Truly independent components of a bound system are incompatible with the system being a single bound entity with properties other than those of a collection of independent parts. Symmetry is always imperfect, however slightly. More drastically, how are bound systems to be modelled if neither those or other simplifications obtain? The answer is found in a statistical treatment. The quantities that we actually observe of an assembly of a large number of constituents are not internal dynamical properties but gross properties, pressure or magnetisation or energy, which in our models are sums or averages over all constituents, and consequently are defined statistically. It is they that we need to calculate. Even so, there are still major problems in statistical physics, as for instance the theory of the equation of state of a liquid. Those problems are problems of our models, for liquids certainly exist, and the statistical distribution of particles within them is both determined by the dynamics and the temperature, and itself determines the equation of state. Nature has solved the problem of the stable constitution of a liquid and so a self-consistent model is there for us to find.

Nature is in major ways uncertain, but we find that in those fields, as in the subjects considered in the previous chapters, theories of uncertain observations are determined by the characters of observations, in particular, by principles of invariance under coordinate transformations and by the fact that observed quantities are collective.

6

Why does mathematical physics work?

6.1 Introduction: the problem

ALMOST ALL, IF NOT ALL, the theories we construct to explain and predict the behaviour of the physical world use mathematics, so much so that to most people, theoretical physics is inseparable from mathematics. The association goes very deep, so much that Ziman (1978) defined physics as *the science devoted to discovering, developing and refining those aspects of reality that are amenable to mathematical analysis*. Mathematics is indeed a very powerful means of establishing the consequences of theories, especially when results are wanted in numerical form, for it is a language free of ambiguity. It is not simply a matter of calculating results. Often the most profound insights into the nature of the physical world and into the structure of relations between different phenomena of physics are disclosed when those relations are expressed in a suitable mathematical way. In the everyday life of a physicist those matters are taken for granted, yet they raise major questions about the nature and purpose of physics.

The questions all relate to the fact that physics is an empirical study whereas mathematics is a logical system. Physics deals with the unique world of nature, the world as it is and no other. Physicists make observations upon that world, observations which of themselves might be no more than curiosities, but it is the aim of physics to put those observations into a rational scheme by which sense may be made of them and of the natural world behind them. Nonetheless physics is an empirical study, and theoretical physics must be an open system of knowledge which can always be changed by new observations. Indeed the argument of the previous chapters has been that the very structure of theory is in certain respects determined by the observations that it is possible to make, including the effects of inherent uncertainty in the physical world upon the observations that can be made, upon the inferences that can be drawn from them, and upon the structure of theory. In each of the previous chapters it was seen how the mathematical forms of particular theories are determined by the relevant empirical observations. The argument is now broadened from particular cases and extended to physics generally.

Mathematics deals with logical relations between mathematical objects and has no necessary relation to anything else. It is true that as a matter of history, certain branches of mathematics have been suggested by physical phenomena, but mathematics in no way depends for its validity upon a connexion with the palpable world. Unlike physics, it is a closed system of knowledge, concerned with nothing but the relations between a defined set of objects.

How then can it be, many have asked, that there is such a close and productive correspondence between mathematics and physics? Why should classical dynamics, or electromagnetism or solid state physics, correspond to mathematical relations? How can we devise mathematical representations of uncertain behaviour? Is it because we somehow select only those phenomena that can be described mathematically and filter out everything else, or is it that the natural world really is mathematical in its essence? Why is it that the branches of mathematics most fruitful for theoretical physics are those apparently most remote from ordinary day-to-day life? It seems on the face of it unlikely that mathematics should be so useful in organising our knowledge of physics, yet from at least the time of Newton onwards the application of mathematics to physics has been quite extraordinarily productive.

We must first look more closely at the uses of mathematics in physics. Mathematics had its origins in essentially practical matters, the need to count in commerce, to measure areas, to tell time, to record the positions of heavenly bodies. Mathematics still finds its most numerous applications in commerce, industry and administration and many uses of mathematics in science are purely descriptive extensions of the original applications. Thus for example, someone observes the pressure of a gas in a vessel at different temperatures. She might just write down all the separate results, but finds that near enough they follow the rule, $p = kT$ (Henry's law), so instead of making a list of values of temperature and pressure, she could equally well represent the observations by the values of just the parameter k and the pressure at some one temperature. The formula is just as good as the list of observations, especially if some uncertainty is assigned to k; it is a great deal more economical, and it may well be more illuminating. Such a formula, a representation of the observations, does not constitute a theory in the strict sense, for a theory should give some justification for the observed value of k, but it does show more clearly than a mere collection of the original data just what a theory has to explain.

While not strictly a theory, the mathematical representation of observations by a formula does itself present a problem, one to which we shall return in the next chapter – why is it that data can be represented by formulae with very small numbers of disposable parameters, often a very minimum number? It is a question that Jeffreys and Wrinch illuminated many years ago (see Jeffreys, 1973; Jeffreys and Wrinch, 1921). They argued that possible representations of a given set of data could be ranked according to the number of disposable parameters. The more parameters the better would the representation agree with the data, but as the number was increased, the improvement would eventually not be significant. The most economical representations are also the most powerful because they can embrace the largest range of observations – the more numerous the parameters the more restricted that data to which they can apply. I do not pursue the matter further in this chapter but it is an important issue in the logic of uncertain argument and the fact that many data can be represented by far fewer parameters would seem to indicate some underlying regularity in nature, or more strictly, in our observations of nature.

One of the most famous representations of physical phenomena by mathematical formulae was the Ptolemaic description of the motions of the planets in terms of epicycles, or combinations of circular orbits. The positions of the planets in the sky as seen from the Earth were derived from their motions in circular orbits, the centres of which themselves moved around circular orbits having the Earth as centre. The scheme represented the past motions of the planets very well, it predicted future positions quite accurately and generally fitted observation much better than the heliocentric model of Copernicus. The heliocentric system remained inferior to the Ptolemaic until Kepler introduced elliptic orbits. Neither the Ptolemaic nor the Keplerian scheme was truly a theory. They were geometrical models that reproduced the positions of the planets in the sky but they offered no physical explanation for the motions of the planets and so the discussion of their relative merits was confined to how well they did reproduce the positions of the planets. The Ptolemaic scheme only succumbed when Newton developed the theory of heliocentric orbits based on universal gravitational attraction.

A true theory is not just a compact description of phenomena, it is a model of the physics involving logical deduction of the consequences of a set of physical postulates – of inertia and gravitation in the case of Newton's theory of the motions of the planets. A theory in this sense will not simply summarise what has been observed in the past, it will predict the results of observations not yet made. It will predict not only the results of observations like those made in the past, it will also suggest other types of observation that might be made. An outstanding instance is Maxwell's theory of electromagnetism: it was a consequence of his equations of the electromagnetic field that electromagnetic radiation should be propagated in free space with the speed of light, as it was later found to be by Hertz. Fermi's prediction of the neutrino, based on laws of conservation, is another striking instance, as is the prediction of gravitational waves as a consequence of general relativity, although as yet they have not been detected by terrestrial observations. A good mathematical summary of data may also predict effectively the results of observations of the same type, but will not suggest the possibility of observations of a different type.

Examples of the use of summaries of data are common in other sciences than physics. Thus, the idea of pattern recognition has been applied in geology and geophysics to make predictions in the absence of deeper understanding. It has been used especially in exploration for oil through the correlation of patterns of seismic reflections, as well as in the prediction of earthquakes. Similar phenomenological arguments appear also in biology. The whole point of natural philosophy is to understand why things are as they are and not simply to describe the world as it appears. The distinctions I have made here are closely related to the topic of 'saving the appearances' – are mathematical theories just instruments for calculating satisfactory expressions for the results of observations, or do they in some way reveal the reality behind the physical appearance? the issues go back to Greek science (Lloyd, 1991).

In this chapter I am concerned with true theories. It seems unlikely that mathematial theories would be as successful as they are, were there not some close relation between the structure of the physical world and the structure of mathematics. Is that indeed so? Certainly it is a question that has attracted over the centuries the attention of many physicists and mathematicians and philosophers, and has generated a very great deal of metaphysical discussion over such issues as 'The harmonies of nature' (Pythagoras) or 'God is a mathematician' or 'God does not play games of chance with us' and so on.

There is no metaphysics in this chapter. I develop a different approach which follows on and extends the previous chapters on metrology and the forms of physical laws. Why does mathematical physics work? Roughly speaking there have been two views, that the world really is a mathematical structure, a collection of mathematical objects; or that we select for study only those features of the natural world that are susceptible to mathematical expression (see Ziman, 1978). The first view has often been associated with a deist view of creation and in some forms at least does seem to imply the notion of a rational creator. The second view, on the other hand, by its emphasis on the human role in the selection of phenomena, tends to a highly subjective view of physics as a purely human construction. I put forward here an argument that is more direct than metaphysics, but at the same time having its own subtle features.

6.2 Abstract groups and concrete realisations

The theory of the vibrational energy of molecules depends heavily on ideas of symmetry. The ionic framework can be treated to a good approximation as a classical mechanical system of masses and springs with normal modes of vibration, and the classification of the normal modes is a consequence of the symmetry of the ionic structure. Every molecule has the symmetry of some point group, that is to say, there is a fixed point in the molecule and a set of rotations about that point that moves the molecule into itself. A diatomic molecule such as O_2 in which the ions are identical, is symmetrical with respect to rotations by π about the mid-point of the internuclear axis or, equivalently, with respect to reflexions in the plane perpendicular to that axis through the mid-point. The molecule is also symmetrical with respect to a rotation of any amount around the internuclear axis, and although that does not affect the vibrational energies, it is a factor that determines the wavefunction of the electrons.

Symmetry means that a structure is unaltered by certain operations such as rotation about some axis through some angle or a reflexion in some plane. If the structure is unaltered, the normal modes of a vibrating system will not change. Thus, if the symmetry operation corresponds to some matrix, that matrix, when it operates on the set of mathematial functions that represent the normal modes, will transform them into themselves. Thus the sets of matrices set out in Chapter 4 that correspond to rotations in 3-space are a representation of the symmetry group of rotations and the functions that correspond to the normal modes are the basis of the representation. That is a very powerful concept because, as already indicated, it allows many properties of the vibrations of the molecule to be derived from the point group of the symmetry of the molecule.

The argument goes deeper. It is not just that the mathematical functions corresponding to the normal modes are unchanged by a mathematical transformation: the actual physical motions themselves must be such that when the molecule is viewed from equivalent directions, they appear the same. The physical structures and motions of actual molecules, and not just their mathematical models, can therefore be the basis for representations of symmetry groups. I now go on to develop that idea more generally.

I begin with a summary of some well-known ideas about groups and their representations, supplementing those that were set out in Chapters 3 and 4.

A group in mathematics is a collection of mathematical objects which are unspecified except for a rule that determines the way in which they combine. Groups are called abstract because the nature of the objects is left open. Besides the rule of combination, members of a group must satisfy some very general conditions. There must be a unit member which combines with every other member of the group to reproduce that same member. The combination of any two members of the group must also be a member – a group would hardly seem to be a group if it were possible to leave it by operations within it. There must also be an inverse of each member such that any member and its inverse combine to give the unit member.

An important idea in physical applications is that of sub-groups. A sub-group is a collection of members of a group, less than the whole group, which itself forms a group. Another idea of physical importance is commutation. A group is said to be commutative (Abelian) if the results of combining members of the group do not depend on the order in which the members are taken, and it is said to be non-commutative (non-Abelian) if the order does affect the result. Many of the groups that correspond to operations in quantum mechanics are non-commutative. A third distinction is whether a group has a finite number of elements or an infinite number. A group that corresponds to a reflexion by π in a single plane will have just three members, the unit member, the reflexion by π and its inverse; it is a finite group. On the other hand a group corresponding to a rotation through an arbitrary angle about some axis will have an infinite number of elements, corresponding to the angles between 0 and 2π; it is a continuous group. All those properties are determined by the rule of combination of the group, irrespective of the precise nature of the objects that constitute the members.

While much can be learnt about groups without specifying the nature of the elements, there are also many purposes for which some explicit form of element must be taken. Any set of specific objects with the same rule of combination as an abstract group constitutes a *representation* of the group. Much the commonest representations are sets of matrices, and examples of a finite and a continuous group now follow.

Consider the operation of reflexion in the y–z plane in three-dimensional space. The x-coordinate of a point will become $-x$, while the y and z coordinates will be unchanged. Thus we may represent the operation of reflexion by the matrix $\mathrm{diag}\{-1,1,1\}$ acting on a vector of coordinates. The inverse is clearly the same matrix, for operating with the one after the other gives the unit diagonal matrix. Including the unit diagonal matrix, the representation of that special reflexion group has three elements. As a more complex example, consider reflexions in all three coordinate planes. The matrices corresponding to individual reflexions are $\mathrm{diag}\{-1,1,1\}$, $\mathrm{diag}\{1,-1,1\}$ and $\mathrm{diag}\{1,1,-1\}$. Again the matrices are their own inverses and they commute. Combining reflexions about two planes gives the three matrices $\mathrm{diag}\{-1,-1,1\}$, $\mathrm{diag}\{-1,1,-1\}$ and $\mathrm{diag}\{1,-1,-1\}$, while combining all three gives the single matrix $\mathrm{diag}\{-1,-1,-1\}$. The representation has eight elements including the unit matrix.

This example shows that a representation in terms of operators must involve a set of objects upon which the operators act, in this instance, vectors of coordinates. The vectors are the basis of the representation. Another example is provided by molecular vibrations where the normal modes of vibration are the basis set.

A simple example of a continuous group is that corresponding to rotation about an axis, for instance, the z-axis. If the angle is θ, the transformed values of the x- and y-coordinates are

$$x' = x\cos\theta + y\sin\theta, \quad y' = -x\sin\theta + y\cos\theta.$$

The group is continuous because the values that the angle θ can take between 0 and 2π are continuous.

The matrices we have just considered give the transformations of Cartesian coordinates. If other coordinate functions were taken to define the position of a point in space, the reflexion and rotation matrices would also be different. The way in which a group is represented by a set of matrices therefore depends on the form of the vectors on which the matrices operate, that is on the basis of the representation. As indicated above, representations do not have to be sets of matrices and bases do not have to be sets of vectors, although they are much the most common.

A well-known example of representations of the same group by different operators on different bases is that of the rotation group in 3-space. The familiar representation is by 3×3 orthogonal unitary matrices operating on Cartesian vectors, but there is also a spinor representation by sets of 2×2 matrices, the Pauli matrices, operating on spinors. The representation on a Cartesian vector basis is

$$\begin{Bmatrix} x' \\ y' \\ z' \end{Bmatrix} = \begin{Bmatrix} \cos\alpha & \sin\alpha & . \\ -\sin\alpha & \cos\alpha & . \\ . & . & 1 \end{Bmatrix} . \begin{Bmatrix} \cos\beta & . & \sin\beta \\ . & 1 & . \\ -\sin\beta & . & \cos\beta \end{Bmatrix} . \begin{Bmatrix} 1 & . & . \\ . & \cos\gamma & \sin\gamma \\ . & -\sin\gamma & \cos\gamma \end{Bmatrix} . \begin{Bmatrix} x \\ y \\ z \end{Bmatrix} .$$

The corresponding representation in which x and y are components of a spinor and not of a vector, is

$$\begin{Bmatrix} x' \\ y' \end{Bmatrix} = \begin{Bmatrix} e^{-i\alpha/2} & . \\ . & e^{i\alpha/2} \end{Bmatrix} . \begin{Bmatrix} \cos\frac{1}{2}\beta & -\sin\frac{1}{2}\beta \\ \sin\frac{1}{2}\beta & \cos\frac{1}{2}\beta \end{Bmatrix} . \begin{Bmatrix} \cos\frac{1}{2}\gamma & -i\sin\frac{1}{2}\gamma \\ i\sin\frac{1}{2}\gamma & \cos\frac{1}{2}\gamma \end{Bmatrix} . \begin{Bmatrix} x \\ y \end{Bmatrix}$$

(see for example, Jeffreys and Jeffreys, 1972 and Chapter 5 above).

The essence of a representation is that the objects of which it is constructed should have the same relations between themselves as do the members of the parent abstract group. Representations are most commonly sets of mathematical objects such as the matrices just set out and the vectors on which they operate, but they do not have to be. Indeed an obvious representation of a symmetry group consists of the physical transformations of appropriate objects, as when we take a cube and rotate it physically about the symmetry axes or move around it to view it from different directions. The operations of rotating the cube or moving around it can easily be seen to satisfy group properties in general, for there is a unit operation (leave everything as it is) and there are inverse operations, while the operations combine in the ways specified by the symmetry group of the cube.

We have already seen that point symmetry groups may be

represented by operations on the actual physical normal modes of vibration of molecules and we now go on to develop that idea quite generally.

6.3 Physical representations of groups

Group theory is useful for understanding and classifying the vibrations of molecules because all representations of a particular group have certain properties in common and, in addition, all representations of a certain class have common properties. Thus if a particular physical system can be shown to represent some group, some of its formal properties will be the same as the formal properties of all representations of that group, while other properties may correspond to those of some restricted set of representations of the group. It follows that once the group represented by a physical system is established, together with the class of representation of that group, the properties of the physical system can be derived from the mathematical models of the same class.

Another way of looking at the idea of representation is through the concept of groups acting on vectors over some field. Many physical quantities can be represented by vectors in an appropriate space, and many operations of physics can be regarded as transforming or mapping a vector of the space into another vector of the same space – an automorphism. If we are dealing with linear physics, operations on the vectors will be linear, and the group with which we are concerned is the general linear (GL) group. We shall assert that a great deal of linear physics can be represented by examples of the GL group.

An example of those general observations is provided by the symmetry of atomic wavefunctions. An atom as a whole is symmetrical under any arbitrary rotation in 3-space, but its properties cannot be described by just a single vector of three components. Many more coordinates and momenta than three are required for most atoms and all will change when the atom as a whole is rotated. The wavefunction is likewise a set of many components and is a vector in a multi-dimensional space. We may think of the effect of rotating the atom as a whole in 3-space as corresponding to a rotation in a many dimensional space of the components of the wavefunction. We may thus define the dimensions of a representation, namely the number of basis functions (independent wavefunctions in this case) needed to describe the system. When the dimension of the representation is known we then know the degeneracy of the atomic system and consequently we know for example the number of states into which a degenerate state will split when some perturbation is applied that destroys the inherent symmetry of the atom. We have already made use of those ideas in Chapter 5.

Two ideas of the theory of representations are important in physical applications. One is that of irreducible representations. When a set of linear operators represents a group, and operates on a set of basis functions, it transforms those functions into linear combinations of themselves, for example by such operations as

$$x' = x\cos\theta + y\sin\theta,$$
$$y' = -x\sin\theta + y\cos\theta,$$

x and y are transformed into linear combinations, x' and y' of themselves.

It may be that the basis functions can be divided into subsets such that members of one subset transform into each other and not into members of another subset; we might for example, take as basis functions the coordinates of two independent points instead of a single one, in which cases the transformations would be

$$x_1' = x_1\cos\theta + y_1\sin\theta,$$
$$y_1' = -x_1\sin\theta + y_1\cos\theta.$$

$$x_2' = x_2\cos\theta + y_2\sin\theta,$$
$$y_2' = -x_2\sin\theta + y_2\cos\theta.$$

The overall representation is composed of two which cannot be further separated. Representations in which the basis functions transform into each other and cannot be divided into subsets are called *irreducible*. They are of great importance in physics, for example, if the basis functions constitute normal modes of vibration of a mechanical system, they all correspond to the same frequency. In a similar way, if the basis functions are eigenfunctions of some operator, then they all correspond to the same eigenvalue. Identification of the irreducible representations of a group therefore establishes much about the observable properties of a system, such as frequencies of vibration, energy or angular momentum.

The other important concept is that of the character of the operator representing some group element and in particular, the identity. The character is the trace of the matrix representative, that is, the sum of the diagonal elements. For example, the trace of the matrix representing rotation of a vector in 3-space through an angle θ about the z-axis is $(2\cos\theta+1)$, and that of the identity element is 3. Similarly, the trace of the matrix representing the same rotation in 2-spinor space is $2\cos\frac{1}{2}\theta$ and the trace of the identity element is 2. These values are examples of a general result that the character of the identity element is equal to the dimension of the representation, that is the number of independent basis functions.

It may be seen from these observations that the degeneracy of a quantum state, or the number of eigenfunctions associated with a given eigenvalue, is equivalent to the character of the identity element of an irreducible representation. Accordingly group theory not only tells us important things about unperturbed degenerate states but it also tells us what the effects of perturbations will be upon the degenerate state. If, for example, we know the group theoretical description of some atomic state, we also know how the state will split when a steady magnetic field is applied to the atom.

Much of the discussion so far has been in terms of the effects of symmetry and how they may be expressed in terms of group theory. Effects of symmetry are of course real physical effects. There is another major field in which group theoretical ideas are important which is more concerned with the way in which we describe physical systems. We saw in Chapter 3 that the formalism of special relativity is a consequence of the measurements that we can make to distant events. The formalism is concerned with the ways in which we describe those events, the space-time coordinates we assign to them or the electromagnetic potentials to which they are subject. Coordinates and potentials are ours to choose, but in such a way that the measured quantities are not altered. Any transformation of the space-time coordinates of events must preserve the space-time interval between the events.

Physics is concerned with the structure of systems, with processes in systems and with the ways in which we ascribe coordinates and other parameters. Structures usually have some symmetry that can be described in terms of finite groups, while processes (especially evolutions of systems in time) and transformations of coordinates normally involve continuous groups. In the next sections, we look at those main types of group in more detail.

6.4 Symmetry: finite groups

Two principal classes of symmetry are encountered in physics. In one, some point in space is not moved and the operations belong to the so-called point groups. The application of that type of symmetry to molecular vibrations and other normal mode problems has already been indicated. The other type of symmetry comprises the space groups in which the operations involve translations. A familiar example in two dimensions is provided by the repetitive patterns of wallpaper, while much of solid state physics turns upon the symmetry of crystals under translations. As with point groups, the nature of a space group determines such properties as the degeneracy of wavefunctions.

Symmetry is never perfect but is always broken by some physical effect. An obvious example is that the translational symmetry of any crystal is defective by reason of its finite size, which means that no translation, however small, actually carries the crystal structure exactly into itself. We have already seen that the point symmetry of an atom is broken when the atom is placed in a magnetic field. We saw above that the maximum number of sub-states into which a degenerate state may split when perturbed is determined by the characters of the representation of the symmetry group to which the physical state corresponds. However, not all the sub-states may appear, for if the perturbing effect has itself some degree of symmetry (for example, if the surface of a crystal is normal to some crystal axis or when the application of a uniform magnetic field defines an axis of symmetry) then the result of the perturbation will depend upon the symmetry of the perturbation as well as on that of the unperturbed state.

The discussion of symmetry and its significance in physics might give the impression that we can observe the symmetry of an atomic or a molecular structure directly just as we can the symmetry of a building. That is not strictly so. We cannot observe directly the normal modes of vibration of molecules, nor phonons in solids nor atomic wavefunctions. The only observable quantities are frequencies, polarisations and intensities of radiations corresponding to differences of energy levels between stable states. The purpose of the whole apparatus of the group theory of symmetry is to understand the structure of those differences and we have already seen in Chapter 4 how considerations of symmetry and group representations in the structure of atoms and molecules led to the conclusion that the correct geometrical framework for microphysics is special relativity.

Symmetry under transformations in three-dimensional space is not the only form of symmetry encountered in physics. In the theory of nuclear structure and high energy physics we encounter effects that are the same whether the particles have positive or negative charge – we say that the system is symmetrical under charge reversal. The formalism of symmetry therefore has very wide applications in physics, for almost any property that is constant in some physical system or process can be related to a symmetry transformation.

The classes of symmetry that we have considered so far mostly correspond to finite groups, for the number of distinct operations that effect the transformations of a particular point or space group can be counted. There are also transformations that are continuous; we now look at some of them.

6.5 Continuous groups: metrology

It has been emphasised more than once that distant events can only be observed with electromagnetic radiation and that however the distant events are interpreted or described, whatever coordinates may be assigned to them, the observed quantities, essentially frequencies and combinations of times of electromagnetic signals, must be kept invariant. In particular the space-time interval must remain invariant under any change of coordinates. Space-time coordinates must therefore transform according to a Lorentz transformation.

Lorentz transformations are representations of the Poincaré

group and the basis is constituted by space-time coordinates. Those coordinates are not directly observable, and indeed only exist in models of the distant events, they do not necessarily correspond to any physical state but are a convenience for calculations about the events. In the same way, rotations of coordinates in 3-space preserve the length of a line element. Whether in 3-space or 4-space, rotations can be of any magnitude and thus there is an infinite set of transformations that represent the Lorentz group. The basis sets of the transformations may be thought of as vectors in a space for which a scalar product is defined and the transformations leave the scalar product of any two vectors unaltered. In particular the scalar product of a vector with itself is a constant; the line element is such a scalar product. We have already seen, in Chapter 3, that metrological conditions require that the rest energy, E_0, the magnitude of the energy–momentum 4-vector, should be a constant, as should the scalar product corresponding to the generalised electromotive force, $(d\mathbf{s}\cdot\mathbf{A})$.

The ways in which we represent the physical world must be such that quantities that are actually measured must remain constant when we change the representation. In particular, if we assign coordinates to events, or potentials to electromagnetic fields, whatever transformations that we may envisage for them must be such as to keep constant the consequential values of measured quantities. That requirement entails certain group structures for large parts of mathematical physics.

6.6 Dynamics

Dynamics is the study of how a physical system evolves in time and the equations describing how a system changes involve the derivatives of various mathematical objects with respect to time. Operations of dynamics usually satisfy semi-group conditions: there is a unit operator for $t = 0$ and a dynamical change over one interval of time followed by one over a second interval gives another dynamical change, but there may not be an inverse operation, especially if the equations of motion are non-linear. The operators of linear dynamics often possess inverses and then form groups. When time is a continuous parameter of a system the group is a Lie group and the behaviour for all time can be derived from the tangent operators at the origin, that is, the differential coefficients at $t = 0$. When the dynamics is non-linear different mathematical structures apply.

Quantum mechanics asserts that the entire observable behaviour of a dynamical system can be derived from a wavefunction and we saw in Chapter 2 how the experimental definition of a standard of frequency leads to the time-evolution equation for the wave-function, an equation which it is evident entails a Lie group structure for quantum mechanics, for

$$\Psi(t+\delta t) = \left(1+\frac{\mathrm{i}}{\hbar}H\delta t\right)\Psi(t)$$

$\mathrm{i}H/\hbar$ is the tangent operator of the Lie group at the origin of time and its algebra determines the behaviour of the system throughout the time. Because all wavefunctions must be orthonormal it follows at once that H is Hermitian.

We have here again much the same situation as with the transformation of coordinates: once we have decided to represent a physical system in a particular way, by coordinates or by a wavefunction, then the metrological conditions, an observed space-time interval or a frequency standard, entail a particular group structure for the mathematics.

Equations of motion that represent a Lie group have the general form

$$(x+\mathrm{d}x) = \Omega\cdot x,$$

where Ω is some operator equal to the unit operator for $\mathrm{d}t = 0$ and generally equal to $(1+A\cdot\mathrm{d}t)$ with A equal to $\partial\Omega/\partial t$ at the origin.

It is easily seen that the familiar equations of dynamics are not of Lie form. Newton's form of the equation of motion is

$$\partial\mathbf{p}/\partial t = \mathbf{F},$$

and other forms of the equations of motion, in particular the Lagrangian and the Hamiltonian do not as a matter of general form involve operators operating on some initial function.

There are certain special cases in which the equations of motion can be put into Lie form, in particular when the Hamiltonian is quadratic in the momenta and coordinates:

$$2H = aq^2+bp^2.$$

Then

$$\partial p/\partial t = aq \quad \text{and} \quad \partial q/\partial t = -bp.$$

We take p and q to be components of a vector. The vector of p and q at $(t+\mathrm{d}t)$ is given by

$$\begin{pmatrix} p \\ q \end{pmatrix}_{t+\delta t} = \begin{pmatrix} p \\ q \end{pmatrix}_t + \frac{\partial}{\partial t}\begin{pmatrix} p \\ q \end{pmatrix}_t \cdot \delta t$$

that is

$$\begin{pmatrix} p \\ q \end{pmatrix}_{t+\delta t} = \left\{1 + \begin{pmatrix} \cdot & a \\ -b & \cdot \end{pmatrix} \delta t\right\} \begin{pmatrix} p \\ q \end{pmatrix}_t,$$

which is manifestly of Lie group form; when the interval of time is not small, we develop the tangent operator as the argument of an exponential,

$$\begin{pmatrix} p \\ q \end{pmatrix}_{t+\delta t} = \exp(M \cdot \Delta t) \begin{pmatrix} p \\ q \end{pmatrix}_t,$$

where $M = \begin{pmatrix} \cdot & a \\ -b & \cdot \end{pmatrix}$.

We may easily see that p and q vary harmonically with time. The rates of change of p and q are found by operating on them with the operator M, that is

$$\frac{\partial}{\partial t}\begin{pmatrix} p \\ q \end{pmatrix} = M \begin{pmatrix} p \\ q \end{pmatrix},$$

whence

$$\frac{\partial^2}{\partial t^2}\begin{pmatrix} p \\ q \end{pmatrix} = M^2 \begin{pmatrix} p \\ q \end{pmatrix},$$

where

$$M^2 = -ab \begin{pmatrix} 1 & 0 \\ 0 & 1 \end{pmatrix}.$$

Thus p and q vary with time as $\exp\{\mathrm{i}(ab)^{\frac{1}{2}}t\}$.

In classical mechanics the Hamiltonian is usually, but not always, quadratic in the momenta, but the potential is often not quadratic in the coordinates.

While the more familiar equations of motion of classical mechanics are not of Lie group form, the derivative with respect to time of any dynamical variable is equal to the Poisson bracket of that variable with the Hamiltonian, in just the same way as the derivative with respect to time of an operator in quantum mechanics is equal to the commutator of that operator with the Hamiltonian. In each case, the operations of time evolution constitute a Lie group with the time interval as the continuous variable and the nature of the evolution in time determined by the algebra of the Poisson brackets or commutators. All Hamiltonian dynamics may therefore be regarded as representations of appropriate Lie groups.

6.7 Physics and group representations

We have now looked at some examples of how a physical system itself does not change with the way we look at it – we say it is symmetrical in some respect – and we have looked at examples of how the ways in which we describe physical systems have to transform so as to leave unchanged the values of observations calculated from them. In both instances, the actual or hypothetical operations of physics can be regarded as representatives of abstract groups and they are so because of the ways in which we observe nature. The correspondence between an abstract group and a physical state or process that is a representation of it derives from what it is possible to observe of that state or process. The natural world may not be a mathematical construct but the observations that we can make of it do correspond to mathematical objects. In that sense they impose a selection, which may be said to be determined by the fact that we as observers are restricted in various ways, as for example to our small local region of all space and time.

We have also seen that classical dynamics can be regarded as representing abstract groups and we now ask if all physics can be put in similar terms. A very large part of theoretical physics certainly can be put in terms of group representations, that is to say, of groups acting on vectors. We assert that we can identify physical quantities that behave as vectors, and physical relations between them that behave as operators, and that provided the relations are linear, then we can consider the physics as an example of the GL group acting on vectors over some defined field.

The vast corpus of exact solutions of field theory in terms of special orthogonal functions that take constant values on orthogonal coordinate surfaces depends on two principal applications of the theory of symmetry. In the first place, the techniques available for constructing solutions depend almost wholly on separating the partial differential equations satisfied by the field variables into ordinary differential equations each of which involves one spatial variable alone. Whether or not that can be done depends upon symmetry operators related to the partial differential operators of the field equations. Secondly, the construction of the functions depends on their being the basis functions for representations of certain symmetry groups (simple examples are expansion of a field variable in Fourier components in a rectangular box, or the symmetry properties of the solid harmonics used to construct solutions inside or outside a sphere).

The mathematical functions that enter field theory are of course basis functions for representations, but so are the physical fields themselves. Indeed the argument here is that physical systems are bases for representations constituted by the physical operations upon them; for example, each of the sound fields that can be independently established in some acoustic resonator is one of a set of basis functions for the representation of the symmetry group of the boundary of the resonator and can be regarded as a vector over an appropriate field.

The factor which controls the mathematics used to obtain exact solutions in field theory is the symmetry of the coordinate surfaces on which boundary conditions are given. In so far as real surfaces never coincide exactly with ideal coordinate surfaces, exact solutions do not correspond to real physical situations and allowance has to be made for deviations. Perturbation theory has been developed to deal with that.

The general scheme of perturbation theory in classical or quantum dynamics is that one starts from an Hamiltonian which admits of exact solutions (coordinates and momenta in classical mechanics, a wavefunction and eigenvalues in quantum mechanics) and adds to it terms, supposed small, that bring it to correspond to a real physical state. One then develops expansions of the perturbed coordinates and momenta or wavefunction and eigenvalues, in powers of some parameter of the addition to the Hamiltonian.

An example of perturbation theory in classical dynamics is afforded by the motion of an artificial satellite about the Earth.

Orbits about a spherically symmetrical Earth are of course constant ellipses. The actual Earth is flattened at the poles and so its potential contains a quadrupole term proportional to the flattening in addition to the potential of a uniform sphere. There are various ways of handling that problem, but the neatest and most powerful, introduced by Delaunay in the nineteenth century, is to make a canonical transformation of the coordinates and momenta in such a way that the Hamiltonian, when rewritten in terms of the transformed variables, allows an exact solution. Nowadays there are more powerful ways of carrying out the transformations than those devised by Delaunay, but all have the same properties, that they depend on a small parameter that is in principle continuously variable, and that they represent a group, in fact, a Lie group. All perturbation procedures that involve small continuously variable parameters can in fact be seen as involving operations belonging to some appropriate group.

To summarise so far, we have seen that special relativity, classical and relativistic dynamics, electromagnetism, field theory, perturbation theory and applications of symmetry in physics, can all be regarded as affording physical representations of some abstract group. Some parts of mathematical physics can clearly be seen as mathematical representations of groups corresponding to representations of the same group by physical phenomena. The theories we have considered as examples of that assertion are very general in form, and it is natural to ask whether similar arguments can apply to the more phenomenological parts of physics, such as the rich and diverse phenomena of solid state physics. Can the argument be extended from a great number of particular instances to much wider generality?

It is not sufficient just to say, for instance, that all solid state physics is an inevitable consequence of the group structures of quantum mechanics and symmetry operators, true though that is. A very great deal of empirical data is incorporated into and deduced from the basic theory in actual solid state physics. Why does mathematical theory work so well in those circumstances? Do we in fact select for theoretical studies those parts of physics, or parts of parts of physics, that fit into a framework that we can handle mathematically?

Much of physics consists, as we have said, either of studies of structures of objects or, on the other hand, of processes that change physical objects. Structures are naturally dealt with in a framework of symmetry, while processes are naturally seen as representations of Lie groups. Physical consistency would seem to imply that operations in the physical world satisfy the fundamental group axioms – the existence of a unit element, the existence of an inverse element, and the associative property. Consistency also requires that there should be a definite rule for combination of the physical elements that represent a group. We might argue in fact, that it is because our observations of the world of physics can be placed in a rational consistent framework, that objects and processes in physics can be elements in the representation of some abstract group.

That is a far-reaching assertion, and no doubt needs more searching examination. An important matter is whether the foregoing arguments apply quite generally or whether they may be restricted to linear mathematics. The examples that have been given are all effectively drawn from linear problems and the argument needs to be developed further to see how far they cover non-linear behaviour, chaotic dynamics, solitons, and the like.

The assertion also raises the question, whether the physical world really is rational and logically consistent, or whether it is that we somehow or other choose to look only at those observations which can be placed in a rational frame. That issue is closely related to the matter of prediction – if up to the present time we can represent certain observations by a consistent mathematical structure, what grounds have we for thinking that similar observations, as yet unmade, will fit into the same structure in the future? Prediction is at the heart of the matter. If we make certain measurements on steel or concrete, we can predict that one bridge constructed of those materials will collapse, while another will stand up. If we make spectroscopic measurements on certain molecules, we can calculate rather well the thermodynamic behaviour of a gas of those molecules. All engineering and much of physics depends upon being able to make successful predictions. Successful prediction seems to depend on three conditions. One is that there are definite rules for the combination of physical elements or processes, just as there are for the elements of a group. Secondly, the physical world should be stable in time, so that the conditions that held when the initial observations were made will continue to apply in the future for which predictions are made. The third condition is that the original observations should encompass the whole range of possible variation, again something of which no one can be sure until the future observations are made. All three conditions are assumptions, and can only be tested after the predicted event. The history of successful physics is the history of people becoming more and more sure that those conditions do hold in the physical world.

It seems that a far-reaching process of selection is being applied, for science only deals with those phenomena for which the three conditions are reasonably well fulfilled. Yet it is not possible to establish the conditions with certainty; only with something less than certainty can we assert that conditions will be in future as they are now, that there are definite relations between physical phenomena and that parameters in expressions representing the existing observations have a wider validity. Those are the principal topics of Chapter 7. They do not necessarily show why mathematical physics works, but they do recall us to the other use of mathematics in physics, the representation of numerous data by just a few parameters, and to the demonstration by Jeffreys and Wrinch (Jeffreys, 1973) that only a few parameters are indeed all that are usually needed, again an argument of a probabilistic nature.

6.8 Dynamics and semi-groups

It was seen in Section 6.6 that Hamiltonian dynamics, that based on the principle of least action, can be represented by operators of Lie groups. Dynamics, regarded as the study of the evolution of any system in time, is much more general than classical Hamiltonian mechanics and the operators belong in general to semi-groups, not groups. A dynamical process, regarded as a continuous flow, may have an inverse and represent a continuous or Lie group but very many dynamical processes do not.

Consider an operation in which a state at time t_0, $S(t_0)$, say, evolves into a state at time t_1, $S(t_1)$, and then subsequently into the state $S(t_2)$ at time t_2. There is clearly an identity operation when t_1 is the same as t_0 and nothing changes. As soon, however, as we consider states at different times, the case is altered. In particular, let t_2 be the same as t_0, so that the operation from t_1 to t_2 is the inverse of that from t_0 to t_1. Because the result of a chaotic operation is not independent of the starting point, it is not possible to say that the operation $A(t_1, t_2 = t_0)\,(S(t_1) \rightarrow S(t_2))$ will be $S(t_0)$. It is not possible to define an inverse operation in chaotic dynamics because the result of an operation which is formally the inverse of some other will not necessarily return the system to its original state.

There is a further reason why it may not always be possible to obtain so close a relation between mathematical results and observation in chaotic dynamics. Because the solutions of equations and the outcome of experiments depend so critically upon conditions, the mathematical results can often do no more than indicate the sort of behaviour to be expected in practice rather than giving an exact match with it. In addition, random noise in practice may obscure aspects of the mathematical solutions.

The properties of semi-groups, as of groups, depend upon the structure of relations between the mathematical objects that constitute the semi-group or its realisation. Among the important properties of semi-groups are the variation of state variables with time, whether or not equilibrium (fixed-point) states exist and if so whether they are stable in time, and the number of variables to specify realistically the state of a complex system. Those properties are determined by linear or non-linear forms of the interrelations of the variables of the mathematical model and the structure of the connections between them. We saw that mathematical models involving groups succeed because constraints of observation impose a structure upon observations and the relations between them that correspond to a group structure. Similarly the nature of dynamical flows and of the gross observations that we can make upon them impose a structure upon observations that corresponds to a semi-group structure.

Chaotic dynamics and uncertain physics require a different approach from traditional physics. In traditional physics we develop a theory from a very few physical postulates from which we deduce logical consequences in a strict formal system. There is for example, a clear rigorous path from Newton's postulate of universal gravitation to the motion of an artificial satellite about the Earth or from the postulate of elementary electrically charged particles to the structure of atoms and molecules. In those circumstances we may establish correspondences between the representations of groups by physical systems and mathematical structures in which geometrical symmetry has a very important part. We have to proceed in a different way in the study of complex systems for which the group theoretical approach fails and where geometrical symmetry is not significant. The order in dynamical systems is, as Fröhlich (1977) has insisted, a kinematical order, of which the best known example in physics is that of the order of electrons in superconductivity. Biological systems in general lack obvious geometrical order but have a very high degree of coherence which is established through kinematical order as a result of dynamical processes; similar remarks can be made about economic systems. The essence of the behaviour of a complex dynamical system is that it depends on the structure of the network of connections between parts. We concentrate on those aspects of the behaviour that are independent of the physics of the individual connections but that do depend on the formal nature of the connections – whether they are linear or non-linear – and the arrangement of the connections. Abstract mathematical models should represent the properties of complex systems if their structures correspond to those of a concrete system, whether physical or biological or economic.

6.9 Conclusion

The argument of this chapter has been that much physical theory can be put in mathematical terms because many physical systems themselves constitute representations of groups, on account of metrological conditions, or geometrical symmetry, or otherwise. There are consequently correspondences between the physics and mathematical representations of the same groups. Two questions are raised by that assertion; the first was mentioned above, namely, how far is physics (or rather, the results of physical measurements) truly consistent as is required for it to be a representation of some group, which leads on to a discussion of the place of probable argument in physics.

The second question is whether all sets of observations in physics can be considered as representations of some group, and that question has already been answered: there are sets of observations in physics with structures that do not correspond to group structures and for which other types of mathematics must be used. A closely related issue is why it seems to be more difficult to represent biology by abstract mathematical models. Physics is straightforward as compared with biology. It is sometimes put forward as a criticism of the methods of physics that only the simplest phenomena are brought forward for examination, whereas biologists are compelled to look at more complex phenomena. There are notable applications of mathematics in biology, to population dynamics, to structure of genes, to structure of organisms and to measures of the differences of proteins, to give a few prominent examples, but they do not seem to go to the heart of fundamental questions in the same way that mathematical models of physics go to the heart of physics. How for example, might one handle mathematically a theory that would treat of the evolution of enzymes to a state in which very complicated sequences of chemical processes enable those processes to require extremely small activation energies? Many mathematical models in biology seem to be formalisations of pattern recognition rather than deductive predictive theories typical of physics. The fundamental theoretical questions in biology seem related to kinematic order in space and time and to be more akin to problems of non-linear dynamics and formal properties of complex structures rather than to particular details of the biology. If that is so there is some prospect of an approach to problems of biology that will have much in common with current developments in dynamics. It will no doubt raise the same issues as in physics, whether the success of abstract theories is a reflection of order in nature or of the ways in which we select phenomena for study.

7

Probable argument

7.1 Introduction

OUR KNOWLEDGE OF the physical world is uncertain. Chapter 5 dealt with ways of representing the behaviour of systems when the mathematical models cannot be solved exactly, whether because they are chaotic in the classical sense or because technical problems of many bodies are involved. At the end of Chapter 6 we saw that while mathematical representations of physics may often be expected to work well, there is never exact correspondence between mathematical models and physical phenomena. That raised the question of how far any correspondence could be extrapolated from the known present into the unknown future. Those problems are brought together in this chapter and we enquire how to investigate uncertain matters and consider what classes of phenomena are susceptible of rational investigation. We also discuss the significance of prediction and how far the success of prediction may be an argument for correspondence between our body of physical theory and the structure of the real world.

It is essential to distinguish two reasons for uncertainty. One is the inadequacy of our mathematical procedures. We are unable to construct mathematical models with exact solutions of the motions of all bodies in the solar system, or of all particles in a molecule or crystal or liquid. That does not mean that the physics is insoluble, nature has solved the equations of motion that we find intractable and the motions of the planets are quite definite, the energy levels of molecules and crystals are in practice well-defined. The failings are in our capacity to find or solve a satisfactory model. The other reason is more fundamental: some, indeed many phenomena are truly uncertain, the physics makes them so, whatever model of them we may construct.

We have also to take into account the effect of experimental error upon the correspondence between model and physics. We need to be quite clear about what we understand by uncertain knowledge. From one point of view we have entirely certain knowledge: we make observations which give definite results. One source of uncertainty is that we repeat observations in what we believe to be identical conditions and get different results. We accept that we cannot specify conditions completely and that such variations will occur; often it is satisfactory to treat them as random and essentially irrelevant to the relation of our physical observations to our mathematical model of them. The results of our observations are statistics, mean values, variances and so on, but they are in a sense as definite as the individual observations from which they are constructed.

Besides inherent variability in the physics and random fluctuations in observations, there is a third source of uncertainty in a discordance between the physical circumstances and the construction of the mathematical model. The world of actual empirical observations is more complex than the most elaborate theoretical model so that unsuspected differences occur between them. Those differences are called systematic errors, they are hidden defects of the theory as a complete model of the observations. Theory here does not mean anything deep, no more than a correspondence between mathematical objects with which calculations are done and physical objects that are the subjects and results of observations. Systematic error is difficult to detect and deal with; the effects of the other two sources are variations in results of observations, whereas the third source may give rise to a discrepancy that does not fluctuate around a zero mean and so is not detected when observations are repeated, whilst the average over repeated observations does not tend to zero.

7.2 Basis of probable argument

We would like to have a theory of probable argument that would enable us to handle questions of uncertain inference in the sciences in the same way that rigorous formal mathematics enables us to make unique deductions from postulates. The difficulty that such a theory will encounter, in common with arguments about induction, is that, unlike deduction, it is not possible to give a general statement of its scope. Arguments based on ideas of chance or probability go back to the seventeenth century, if not earlier, and were initially concerned with theories of wagers, annuities, life assurance and the like. The first ideas were combinatorial and were in terms of taking samples from well-defined populations. Somewhat later Bayes and Laplace, still thinking about games of chance, introduced the notion of degree of belief. Ever since there have been broadly speaking two approaches to uncertain inference – do we try to determine statistical chances of obtaining certain samples from some real or hypothetical population, or do we try to set up a mathematics of degrees of belief? Within the latter we should distinguish between subjective probability, the degree of confidence that I actually have in my own mind, and degrees of justified belief, the confidence that we ought to have, given the facts as we know them.

The statistical theory of sampling is very important in the sciences and has been applied to uncertain argument in two ways. It was used as the basis of statistical thermodynamics by Willard Gibbs who saw the actual state of a physical system as a sample from the population of all possible states of that system. More generally, it was the basis for theories of probable inference developed especially by R. A. Fisher and J. Neymann in which any actual state of affairs was regarded as a sample from all possible states of affairs, and Fisher and Neymann introduced statistical objects and criteria that had the aim of giving objective assessments of probabilities of statements without arbitrary assumptions. Likelihood, fiducial probability and confidence intervals are concepts that have become very important in statistics and they were first defined in those studies of Fisher and Neymann.

In the statistical approach to uncertain argument, probabilities are defined as frequencies of events in populations or samples, and that definition is called the frequency definition of probability. There is a fundamental question about it. In such problems as taking samples from bags containing unknown numbers of black and white balls, we do not know the relative numbers of black and white balls, we may not know the total number, but we do know that the parent population is finite and that it is completely described by just two quantities, the total number and the ratio of black to white. The combinatorial calculations are then unambiguous. As soon as we appeal to the idea of an infinite or quasi-infinite parent population, the unambiguous basis for calculation is lost – we do not know what parameters we should use to characterise the parent population. This argument, put very clearly by Jeffreys (1973) and by de Finetti (1974, 1975), shows that the sampling approach is not free of assumptions as Fisher and Neymann had intended, and that it is not one on which a theory of probable argument in physics can be constructed. We also run into difficulties as soon as we admit the possibility that underlying conditions may be changing or that our observations are not representative of the whole range of possible conditions. That is not the same as the problem of infinite populations, although it is a factor in that problem, for underlying conditions could be different in different sub-populations of a finite population. It seems that we have to contemplate infinite sequences of infinite populations to encompass those possibilities.

The alternative to the frequency definition, as argued by Maynard Keynes (1921), (Ramsey, 1926, see Sahlin, 1990), Jeffreys (1973) and de Finetti (1974, 1975), is to set up a theory of degrees of belief, the Bayesian approach, in which the existence of unprovable hypotheses is explicitly recognised, with the consequence that the dependence of conclusions on those hypotheses can be set out. There are differences between the formulations of Ramsey, Keynes, Jeffreys, de Finetti and others, one of which at least is significant, as will be considered below, and the formulation which seems most practically adapted to issues in the physical sciences is that of Jeffreys, who was also the only one to develop a comprehensive theory with applicable algorithms, for the other authors to a greater or less extent confined themselves to general principles. Jeffreys's ideas, which are finding ever increasing applications not only in the sciences but in economics and other domains (see Cook, 1990), are followed in the next sections.

The object of a theory of degrees of belief is to place in order the degrees of belief in alternative propositions about the world or our observations of it, and as Ramsey (1926) emphasised, a principal purpose of that is to guide us in deciding on actions. One of the earliest scientific applications of a theory of probability was in the discussion by Daniel Bernoulli (1760) and d'Alembert of the risk or utility of inoculation against smallpox. The great majority of the applications of probability in physics are to answer such questions as 'are the results of these two experiments in sufficient accord that it is not necessary to make further experiments, or are they so discordant that further work should be done?' That is also the case at a more general level when we assess the validity of a theory. We may put the question, 'is this theory probably true?' but that is equivalent to asking 'is this theory the one most likely to give the correct predictions of the results of observations?'

If a theory of probability is to be useful in physics it must be able to give answers to that class of question. It should be one with which calculations can be done, a mathematical theory, for although the inferences are uncertain, the calculations of their probabilities should be unambiguous. It should also be of general use, that is to say independent of the person making the calculations, because the results should be acceptable to the community of physicists and not just the private belief of an individual. An important difference between the positions of de Finetti and Jeffreys is that de Finetti explicitly states that degrees of belief are subjective and that any assessment is an assessment of the degree of belief of the particular person who makes it, whereas Jeffreys (in accordance with Keynes (1921)) sets up a theory which implies that the results are objective, although he does not seem to state that explicitly. Jeffreys is thus concerned with probabilities as justified degrees of belief. He certainly asserted that the ordering of probabilities was unique, and the whole of his treatment of the estimation and use of probabilities seems to be dependent on the assumption that it is possible to set up objective rules for the manipulation of probabilities, independent of personal assessments of degrees of belief. Ramsey (1926) established a relation between a person's subjective degree of belief in a proposition and the probability to be assigned to it, but for use in science we need a prescription for action that will be agreed by all physicists or all biologists as the case may be. That is why calculations of probabilities should be unambiguous. We are also led to the idea of an operational definition of probability. Just as observations are defined operationally by the way in which they are made, so probabilities are defined for a community of scientists by the way they are calculated. I define probabilities in this chapter by the methods of calculation set out by Jeffreys (1973).

7.3 Mathematical probability

In this section I set out the essential elements of Jeffreys's presentation of the theory of degree of belief as a basis for uncertain inference. It is founded on axioms in much the same way as Russell and Whitehead purported to found pure mathematics on axioms in *Principia Mathematica.*

The first axiom is that the probability of some event or proposition cannot be thought of in isolation, it depends on other knowledge. We can only assess probabilities in relation to specific other knowledge, and to express that idea, Jeffreys introduced the explicit notation for a probability,

$$P(p \mid q),$$

meaning, the probability of the proposition p, given the proposition q.

The background knowledge can often be usefully split up in the form

$$P(p \mid q, H),$$

where q is a particular proposition or propositions and H is general background knowledge.

Jeffreys postulated that probabilities could be ordered uniquely and showed that if so, no inconsistency arose.

That means that if

$$P(p_1 \mid q, H) > P(p_2 \mid q, H)$$

and if

$$P(p_2 \mid q, H) > P(p_3 \mid q, H)$$

then

$$P(p_1 \mid q, H) > P(p_3 \mid q, H).$$

There are similar results for the probability of the same proposition p on different sets of data, q_1, q_2, q_3.

Not everyone who adopted the Bayesian standpoint accepted that the ordering of probabilities was unique. Keynes did not and de Finetti also had a more subjective idea – the ordering might depend on the person who formed the degree of belief.

If the principle of unique ordering is accepted, then there is a one-to-one correspondence between probabilities and the real numbers. Probabilities do not have to be equal to the real numbers, but if we can calculate the real numbers to which they correspond, then we know the order of probabilities. Thus the correspondence with real numbers enables algorithms to be constructed for mathematical calculations with results that can for many purposes be treated as probabilities themselves, although strictly they give the order of probabilities or degrees of belief.

We often want to estimate the degree of belief in two propositions jointly. If H represents the common information and if p and q are the two propositions, the probability of the two together, given H, is written as $P(p, q \mid H)$, and it is derived from the probabilities of the propositions individually by the formulae

$$P(p, q \mid H) = P(p \mid q, H) \cdot P(q \mid H),$$

and equally

$$P(p, q \mid H) = P(q \mid p, H) \cdot P(p \mid H).$$

The significance of those formulae is that the prior information relative to one of the propositions includes the probability of the occurrence of the other, it being supposed, of course, that neither proposition is known to be certain when the joint probability is evaluated.

If p and q are independent, so that the probability of one does not depend on the other, then q is not to be included in the prior information about p, and vice versa. The rule for the joint probability then reduces to the simple product:

$$P(p, q \mid H) = P(q \mid H) \cdot P(p \mid H),$$

the form in which the rule is sometimes stated without qualification. The simple form is correct only if the probability of p on H does not depend on the probability of q on H and conversely.

The product rule in its correct form provides the formal answer to a question that is central in science. Suppose we wish to test the validity of some proposition, some theory. If we have a suitable statement of the theory, we can work out the probability of obtaining certain observations according to the theory. The observations are made; what do they tell us about the degree of belief that we should have in the theory? For example, the theory might predict that a certain quantity should have some particular value. A series of measurements of that quantity has a mean value that does not agree with the predicted value, but the observations have a scatter that we state as a standard deviation. How should we estimate the degree of belief that we should have that the theory represents the physics?

The product rule tells us that the joint probability of getting the observations (q) and of the theory being true (p) is both

$$P(p, q \mid H) = P(p \mid q, H) \cdot P(q \mid H),$$

and

$$P(p, q \mid H) = P(q \mid p, H) \cdot P(p \mid H).$$

The two expressions must be the same and so we have

$$P(p \mid q, H) = \frac{[P(p \mid H) \cdot P(q \mid p, H)]}{P(q \mid H)}.$$

$P(p \mid q, H)$ is of course the quantity we wish to evaluate, the probability that measures our degree of belief in the theory, given the information H and the observations (q).

The probability $P(q \mid H)$ is rather indefinite, since it means the probability of getting the observations in the absence of any theory about them. However, it is often the case that we do not need to worry about its precise meaning, in particular when we want to compare our degrees of belief in two theories (1 and 2) given a set of observations, that is, we want the ratio of $P(p_1 \mid q, H)$ to $P(p_2 \mid q, H)$. Accordingly we may write

$$P(p \mid q, H) = k \cdot P(p \mid H) \cdot P(q \mid p, H),$$

where k is a constant that will be the same for a given set of observations.

The formula incorporates what is known as the principle of *inverse probability*, a direct probability is the probability of observations given a theory, inverse probability gives the degree of belief in a theory following a set of observations.

A common example is that we wish to know whether it is more or less likely that a set of observations depends linearly on some independent variable, or whether they have a quadratic or other dependence of higher power. We might then take p_1 to be the probability of linear dependence and p_2 that of quadratic dependence.

$P(p \mid H)$ is known as the prior probability of p given the information H that was available before the observations were made.

$P(q \mid p, H)$ is the *likelihood* of obtaining the observations given the theory to be tested. The concept was introduced by R. A. Fisher who emphasised the great importance of that mathematical object. It is the term in the expression for $P(p \mid q, H)$ that contains all the information about the observations and consequently Fisher argued that conclusions about relative probabilities should be based on the likelihood alone in order that they should not depend on assumptions that could not be verified. Jeffreys took a different view, one that has essentially been followed ever since by probabilists who accept the Bayesian position, for he considered that it was impossible to avoid assumptions in making estimates of probabilities, and that any such assumptions should be made explicit.

Jeffreys argued that Fisher's procedures could nonetheless be followed in many practical instances. He and Dorothy Wrinch showed (see Jeffreys, 1973) that when the number of observations is large, the prior probability has little influence on the overall probability and can effectively be ignored so that the ratios of probabilities of different hypotheses can be found from the likelihoods alone, as Fisher had proposed. In such cases the prior probability does not have to be assigned. Most frequently the prior probability is not known and so cannot be included in the estimation of probabilities of hypotheses; that has been seen as a fundamental criticism of the Bayesian approach which appears to require some knowledge of prior probabilities. Jeffreys devoted a great deal of space in his book to constructing mathematical forms for prior probabilities that should express the same ignorance of p no matter in what functional form they were written. In this way and in his justification of the use of the likelihood by itself, Jeffreys showed how to get round the need to know the prior probability. In the one case (large numbers of observations) it is irrelevant, in the other case the fact that it is unknown is explicitly included in the mathematics.

The term *prior probability* gives rise to some misapprehension. It does not mean *a priori* probability in the sense of some innate primitive knowledge or logical principles. It is simply a statement of whatever knowledge we had of the state of affairs before starting a particular investigation. It can be quite empirical, for it is the distillation of existing knowledge; it may express some physical or dynamical principles which we believe apply generally. On the other hand it may be a formal statement that we know nothing. It should also include unstated assumptions, and one consequence of writing it down formally is that attention may be called to the fact that such assumptions are being made, as for example, that underlying physical conditions do not change between one set of observations and another. That explicit acknowledgement of underlying assumptions is especially important in the application of probabilistic arguments to chaotic physics. Whenever a prior probability occurs in an expression, it raises a question, and that question must be answered if we are to construct a reliable argument.

7.4 Scientific inference

There is a number of different levels at which scientific argument depends on considerations of probability. At the practical level the problems with which we have to deal may typically be of the form: we have certain data that could imply various alternative propositions; how should we compare our justified rational degrees of belief in those propositions? The answer to that is, by inverse probability. If prior probabilities are irrelevant, that is equivalent to the comparison of likelihoods, but it may not be possible to ignore them. Thus, certain data might be fitted by linear, quadratic or cubic forms and the likelihood of the linear form might be appreciably less than those of the quadratic and cubic, which themselves might not differ very much. If the prior probabilities were irrelevant, the quadratic form might be chosen on grounds of simplicity, but if there were some other consideration that excluded the quadratic form, such as an argument based on symmetry, then the cubic would be selected even though its likelihood were less than that of the quadratic form. If the prior probability of a proposition is zero, the proposition cannot be saved by any observations.

The above remarks mention the argument from simplicity as grounds for choosing one proposition rather than another. The idea that a theoretical model of the physical world should be in some sense as simple as possible goes back a long way; is it just a question of elegance or avoiding trouble, or is there something deeper to it? One answer is to be found in the difference between a mathematical expression as a fit to data and as a deduction from a theoretical model. The point has already been made in relation to expressions for the motions of the planets, but may be repeated here from a rather different point of view. The mathematical forms that express the Ptolemaic geocentric system can be made to work quite well, but only at the expense of continual elaboration. The Newtonian heliocentric scheme, on the other hand, is of a fundamental simplicity that can readily encompass new observations. It should perhaps be said that in reality, the Newtonian scheme is not so simple and the perturbation calculations that are needed to take account of the interactions of the planets are not transparently easy.

The discussion so far begs the question of what we mean by simplicity. It seems that it is at bottom a matter of the number of disposable parameters in a mathematical model (see Chapter 1 and Jeffreys and Wrinch, 1921), for it appears that the more fundamental the model, the richer the results we can derive from arbitrary assumptions and hence the fewer disposable parameters we need. A good theory enables us to organise a very large body of observation on the basis of a very few assumptions, what Leibnitz called the 'principle of plenitude'. If we define simplicity in that way, we can show, as in the following argument, why repeated observations give us confidence in the future.

We suppose that some hypothesis, q, leads to a sequence of observable consequences, $p_1, p_2, p_3, p_4 \ldots$. If the initial information is H, then the joint probability of the hypothesis and the first consequence is

$$P(p_1, q \mid H) = P(q \mid H) \cdot P(p_1 \mid q, H) = P(p_1 \mid H) \cdot P(q \mid p_1, H).$$

But p_1 is a consequence of q so that $P(p_1 \mid q, H) = 1$, and then

$$P(q \mid p_1, H) = \frac{P(q \mid H)}{P(p_1 \mid H)}.$$

Going on now to the second consequence, we replace p_1 by p_2 and H by (p_1, H) (for p_1 is already known to have occurred). Thus

$$P(q \mid p_1, p_2, H) = \frac{P(q \mid p_1, H)}{P(p_2 \mid p_1, H)}$$
$$= \frac{P(q \mid H)}{P(p_1 \mid H) \cdot P(p_2 \mid p_1, H)}.$$

Continuing, we find, when all n consequences have been observed, that

$$P(q \mid p_1, p_2, \ldots p_n, H)$$
$$= \frac{P(q \mid H)}{P(p_1 \mid H) \cdot P(p_2 \mid p_1, H) \ldots P(p_n \mid p_1, p_2, \ldots p_{n-1}, H)}.$$

Of the individual terms on the right side, $P(q \mid H)$ is either zero or positive and less than 1. If it is zero, the probability of q given the observations will always also be zero. On the other hand if $P(q \mid H)$ is not zero, the right hand side will be positive because all the factors in it are positive. The terms in the denominator, being probabilities, are all less than one, but the right side itself is also a probability and cannot exceed 1. That can only be so if each additional factor in the denominator, corresponding to the verification of a further consequence, approaches 1 yet more closely. The chance of successive consequences of q being verified therefore approaches certainty as the number of verified consequences increases. It is this that gives us confidence in the consistency of physics. Since all the factors in the denominator on the right side are less than one, the probability of q given the observations also increases with each verification and so tends to its maximum possible value of 1. The flaw in that argument is of course the unstated assumption that H remains the same in all the terms, that conditions really are unaltered in each test of the hypothesis. That leads us to ask, if successive tests do support the hypothesis q, does that mean that H is unchanging? Yet again, however, we cannot avoid the question, whether the consistency of physics is inherent in the physical world or whether it is constructed by us in our study of that world, including the axioms of our theory of probability.

A related question is whether probabilities, as degrees of belief, are subjective or objective. de Finetti considered them to be subjective, particular to each person who formed them. Keynes also seems to have a similar view, and certainly considered that the ordering of probabilities was not necessarily unique. Jeffreys took a more objective view, although he does not seem to have considered the question explicitly. It seems that an objective theory of probabilities is necessary if physics is to be based on probabilistic argument. My personal degree of belief would not appear to have any necessary relation to the state of the natural world, yet if we are to have a physics that gives a rational account of the world and which leads to verifiable predictions of observations as yet unmade, it would seem necessary for degrees of belief to be related to the natural world and not to me. More strictly, probabilities should be relative to our models of the natural world and not to the world itself, but since our models are not personal to individuals but are common property of science, it still follows that arguments about their validity should be based on probabilities that can be evaluated in some common objective way. The issue is parallel to that raised by the nature of observations. The empirical basis of physics is constituted by the results of observations that the community of physicists accepts as reliable, and it has been emphasised a number of times that we take those results as the subject of theory without trying to go behind them. In a similar way the probabilities of degrees of belief calculated with a communally accepted algorithm have objective validity as empirical subjects of theory on the same basis.

7.5 Sources of uncertainty

Various reasons for uncertainty in physical argument have been mentioned in earlier places, and it is convenient to assemble and summarise them here.

The most common and generally known is the so-called random error, the variation in repeated observations that can be assigned to no definite cause (if it can be so assigned it is called a systematic error). Often such errors arise from imperfections in the means of observation, or from variations in environmental conditions, but sometimes they seem more fundamental. We know well how to handle random errors, how to estimate and express their magnitudes and how to allow for them in deciding whether or not agreement or disagreement between observations is significant. Random errors are a fact of life but they are not relevant to the major issues of the validity of physics as a description of the natural world.

A far more important source of uncertainty is that inherent in the natural world. The mathematical models that we used in physics were until recently linear models, that is to say the mathematical objects that appeared in them were made up of linear combinations of variables, differentials and integrals: powers and products did not appear. A very great deal of our observations of the physical world could be organised in a rational way with such models, which include classical dynamics, electromagnetism and non-relativistic quantum mechanics. The arguments of Chapter 6, in which the effectiveness of mathematics is related to correspondence between physical and mathematical representations of groups, manifestly apply to linear models, they apply also to some non-linear models, and to certain aspects, but only certain aspects, of chaotic dynamics. Linear models, and in particular those just mentioned, deal essentially with single objects, typically with single particles in given fields or a single field. Pairs of particles are also amenable to linear models, but beyond that, one soon encounters significant non-linear interactions. Some linear models are from the start inadequate. Non-linear interactions are one of the causes of chaotic behaviour; they do not always give rise to it and there are other reasons for it as well. Non-linear interactions may have the effect that the state of a system, instead of evolving with time in a regular predictable manner, varies erratically. Geophysics affords many examples. However, there are usually certain stable features of such erratic time series, for example, mean values, variances and autocorrelations, and while they cannot be used to predict values at particular times, they can give overall estimates of the general course of the behaviour.

Chaos also arises from interactions of very large numbers of elements. The classical example is a gas of very many particles all interacting with each other. Each interaction can in principle be calculated by classical dynamics but after just a few collisions the memory of the initial conditions is lost so that the path of each particle is effectively random. In that and similar systems the number of separate particles or sub-systems is so great that the probability of the system overall attaining some particular equilibrium state has a very sharp maximum. That means that average or total properties, such as pressure or magnetisation, have very well-defined values; it is they of course that are observed and not the individual particles. Consequently it is possible to set up models in which the statistical properties follow almost rigorously deterministic laws, as in classical thermodynamics.

It seems quite generally to be the case that there are features of chaotic dynamics that follow deterministic laws and are predictable, as averages or other overall properties; an important aspect of chaotic dynamics would seem to be to identify such predictable properties and relate them to the underlying physics.

Beyond those now reasonably well understood sources of uncertainty in the natural world itself, there are others yet more fundamental to the construction of models of the physical world. The emphasis of this book has been upon the requirement that models of our observation of the world should reproduce the results of observation and upon the idea that the requirement imposes conditions upon models. We have seen also why in such circumstances, models are naturally mathematical. Two doubts lurk behind such assertions – is the natural world stable, in the sense that it is sensible to suppose that the conditions of observation do not change, and how far does the real world, independent of us, conform to the models we make of it? Does the fact that our models are effective in predicting observations tell us anything about the orderliness of that world underlying our observations?

7.6 Predictability and its implications

It is commonly said, and has been said a number of times in this book, that a striking feature of physics is that it can predict phenomena not previously observed. We now want to draw out the epistemological implications of that assertion, but first we must define what we mean by prediction.

One form of prediction is that entailed in Jeffreys's argument (Section 7.4) that repeated observations of consequences of some hypothesis strengthen the degree of belief in that hypothesis, and may do so to such an extent that one is prepared to assert that the next and subsequent trials will also be successful. The next and subsequent phenomena have not occurred when we make the assertion, and in that sense the assertion is a prediction, but it is not a very strong form of prediction, for we are expecting to see phenomena of the same sort as we have seen up to that time.

A stronger form of prediction is made when some wholly new and unexpected type of phenomenon is predicted. An example that is often cited is that of Maxwell's notion that light was a form of electromagnetic wave motion, as implied by the equations for the electromagnetic field vectors that he had established. But what was the basis of Maxwell's equations? In fact the logical train starts earlier, with Faraday's experiments on electromagnetic induction. Those experiments showed that the spatial variation of an electric field over some surface bounded by a conducting wire was proportional to the variation in time of the magnetic field through the same surface. In a similar way, the experiments of Oersted and Ampère linked the spatial variation of a magnetic field to the time variation of an electric field. Those results were empirical. We now see very easily that when the spatial and temporal variations of some quantities are connected in that way, some sort of wave motion must occur, provided we are prepared to think of electric and magnetic fields without the material supports of coils of wire. Nowadays the connection is obvious but like most obvious results, it required deep insight and imagination for Maxwell to see it for the first time.

Another instance of successful prediction, also often quoted and to which reference was made in Chapter 6, is the discovery of the neutrino. Again the starting point was an empirical observation, the spectrum of β-rays from radioactive substances. That appeared to be inconsistent with conservation of energy and angular momentum if an electron was the only particle emitted in the process and Fermi proposed that the resolution of the difficulty lay in the existence of an additional fundamental particle, the neutrino. Here, as with the prediction of electromagnetic waves, the logical consequence of an empirical observation was shown to imply an unsuspected phenomenon. In Maxwell's case, the relation of a spatial variation to a variation in time means that if there is an harmonic variation in time, there must be an harmonic variation in space, but we rely on the consistency of nature when we accept that a relation established with coils of wire on a laboratory bench can also be applied to fields in empty space. Fermi's prediction also depends on the consistency of nature, for we assert that a general dynamical principle, namely that physics is independent of arbitrary origins of time and orientation so that energy and angular momentum are conserved, is as applicable in the interior of a nucleus as anywhere else in nature.

There is a further field of prediction, the extrapolation from laboratory experiments to applications in technology and engineering. The behaviour of a suspension bridge is predicted from tests on steels in a laboratory and the bridge stands up. Again we are faced with the question, if engineering works, can nature be inconsistent? Is the effectiveness of engineering a sound argument for a rational structure to nature?

A successful prediction means that a relation between observations holds on a set of observations that is more extensive than that on which it was first established, whether we think of electromagnetic waves in free space as extending the circumstances of Faraday's laboratory, or the behaviour of steel in a suspension bridge extending the circumstances of a materials testing laboratory, or conditions in an atomic nucleus extending the set of circumstances over which the conservation of energy and angular momentum were first established. Predictions of that sort, far outside the field in which theoretical relations were first established, do not seem consistent with extreme philosophical positions that see the constructs of science as the consequences of sociological factors and so irrelevant to the nature of any independent physical world, if such exists.

Physicists begin with a set of states that may be determined solely by themselves (solipsism, a purely subjective world) or be independent of themselves (naïve realism, a purely objective world), or be some combination of the two (an objective world but with subjective elements). They construct upon that set of states the results of observations that may be represented as vectors in a suitable space. Theoretical physicists devise maps from the space of observations and a space of parameters to a space of theoretical results and then close the circle by a mapping from that space of results back to the space of observations. The specification of a mapping requires not only that a mapping function should be given but also that the space of states should be provided with a topology. If two state sets are mapped to spaces of results and only a mapping function is specified, it must be assumed that the two state spaces have the same topology, in particular, for example, that the norms and scalar products of vectors are formed in the same ways.

Theoretical physicists adjust the mappings and parameters until a good match is obtained to the original data and they may extend the domain of states over which the observations are defined to obtain a mapping of wider application. Prediction involves extending the domain of observations to a set of states distinct from that on which the mapping was first established. A successful prediction means that the same class of observations and the same theoretical mapping as the original investigation remain valid for the new set of states and therefore, by the above argument, that the new set has the same topology as the initial one. That new set was unknown at the time of the original investigation and so could not have been affected by any subjective structure in the original set. The validity of the theoretical organisation of both sets of observations, as established by the successful prediction, thus implies a consistency between the two that is independent of any subjective element in the scientific process. I accordingly assert that a scientific theory has some objective validity and that successful prediction is evidence for order in the world independent of the observer.

We may also adapt the result of Jeffreys and Wrinch (1921) as set out in Section 7.4. We let q be the proposition that there is an independent objective physical world and take $p_1, p_2, \ldots p_i \ldots$ to be the results of theoretical predictions which depend on that proposition. Then given the very large number of successful predictions we conclude that there is such an external independent world.

Prediction implies a consistency in nature, or at least in such of our observations of nature that we designate as physics. Jeffreys's argument of Section 7.4 may indeed be applied to show that the more successful predictions that are made, the more confident can we be in the inherent consistency of observations of nature. At the same time we have to realise that there is never complete certainty in science, neither in the observations that have already been made, nor in the continuation into the future of conditions that have made prediction possible up to now. We deal throughout in probabilities and degrees of belief.

8

Conclusion

8.1 Introduction: the content of observation

MY INTENTION IN THIS BOOK has been to consider how far and in what ways, the formal structure of theoretical physics is determined by the observations that it is possible to make of nature. Before summing up the conclusions of the previous chapters, which are not necessarily comprehensive, I should re-emphasise just what is done in theoretical physics. The basis of physics is the observation of nature, but we must distinguish clearly between the natural world, which is independent of us, and our observations of it, which are not. Theoretical physics is our attempt to give a rational coherent account of our observations and their implications. It may or may not correspond in detail to features of the natural world, for we can do no more than check that our theoretical calculations reproduce results of observation, in particular, the prediction of observations still to be made.

Observation is never an isolated activity. The way that we observe depends on human capabilities and properties of nature. Observation may affect the objects observed and our observational procedures depend upon the state of technology and are guided by theory. The results of observation have to be derived by procedures that depend upon some theoretical model as well as upon experimental techniques, and we define observed quantities operationally by the procedures that we use to obtain them. The harder we question nature, the more fundamental the observations we make, the more dependent are the results on technique and theory. When an astronomer observes celestial objects by electromagnetic radiation, whether at very low or very high frequencies, she will most probably not be at the instrument which will be in a spacecraft or on a remote mountain. The link between the object and her results involves remote control of pointing of a telescope and setting of detectors, the reception of electromagnetic signals, elaborate manipulation of those signals to detect significant effects through noise, and finally transmission by electromagnetic signals for ultimate display as digits on a computer in her own office. Although the astronomer depends upon a whole corps of technicians, it can be said that she by herself controls and operates the remote telescope. In high energy physics similar remote operations are carried out by a substantial team of physicists, themselves dependent on an even larger group of engineers, operators and computer technicians. The investigation of the remote depths of the universe or of interactions at the highest energies requires very expensive equipment so that only a very few examples can be afforded world-wide, and they also entail very complex and lengthy physical and logical links between what it is supposed is being observed and the results that appear in the store and on the screen of a computer. The examples that have been put forward in previous discussions to show that observations have theoretical content (see Chapter 1), seem by comparison almost trivial. All that is true, but it is those final results that the community of physicists accepts as the empirical basis of physics that we have to organise into a logical structure by a theoretical construction; we do not go behind the observational results.

The conclusion is not confined to physics or the physical sciences. Today the fabrication and study of very small and elementary solid state devices involve expensive and elaborate apparatus, methods and analysis. Much chemistry depends on nuclear magnetic resonance, an expensive and by no means transparent technique. Biologists employ synchrotron radiation, the product of high energy electron accelerators to study complex molecules and living objects in the brief instant before they are destroyed by the radiation. The identification of sections of a strand of DNA involves complex manipulations and a considerable basis of theory. Modern science affords us, wherever we look, forceful examples of complexity of technique and theory in the generation of observations or experimental results. In some cases, in very high energy physics or in the breeding of transgenic animals, it is by no means clear whether nature is just being probed or whether it is being constructed.

8.2 Measurement, observation and theory

The very definition of a standard of measurement gives it certain properties in consequence of the fact that once a way of realising a standard has been defined, that realisation is invariant – there is nothing else against which it can be compared or with respect to which changes in it can be established. Thus the standard of frequency is invariant in time and so mathematically the signal by which it is realised is represented by a single Fourier component, with the time variable equal to the inverse of the standard frequency. The notion of a Fourier component is of course a concept of theory. We then adopt the fundamental postulate of quantum theory, that all the observable information about a physical system can be contained in a wavefunction, in the sense that appropriate operations on the wavefunction will yield the results of any possible observations. We further suppose that eigenvalues of wavefunctions exist and that the signal by which the standard of frequency is realised corresponds to a transition between two such states. Wavefunctions, eigenstates and transitions between eigenstates are all theoretical constructs, as is the idea of an operator that brings about a transition. If we let the standard radiofrequency signal, represented theoretically by a single Fourier component, correspond to a quantum mechanical operator of the theory, we obtain the time-evolution equation for the wavefunctions of the two eigenstates postulated in the theory.

The theory of special relativity handles the consequences of the fact that the only observations we can make of events across empty space are instants, frequencies and polarisations of electromagnetic signals. Times, coordinates, momenta, energy, electromagnetic field vectors or tensors, are all variables of the theory used to calculate results of observation, which are essentially, times and intervals equivalent to distances, speeds and accelerations. How we assign them is determined by two factors, the observations that we can make, which must not vary with changes in the way we make the assignments, and the fact that four numbers (coordinates of time and space) are needed to locate any observable event, four components of a vector to specify the dynamics and four components of a further vector to specify an electromagnetic field.

The observables of microphysics are the frequencies and intensities of electromagnetic radiation emitted from or absorbed by a physical system undergoing some change. The interior physics of an atom or molecule is unobservable and in a strict sense unknowable, but we suppose that a theoretical model of that physics should be placed in the same four-dimensional geometry as observations of distant events on the large scale. In so far as there is a necessary connection between the two scales it lies in the assumption that the forces between components of microscopic systems are electromagnetic. Changes of force between charged particles would then propagate with the speed of light and electromagnetic interactions should be incorporated by adding the 4-vector of potentials to the 4-vector of energy and momentum. All that however is a model that cannot be tested directly but is justified by the successful calculation of the observations – the frequencies and intensities of certain electromagnetic signals.

It is very important to keep clearly in mind what are essentially auxiliary objects of the theory and what are real observables. It is easy to think of coordinates as real quantities, but they are only assigned for reference and to help us to calculate real objects, angles and distances. Map makers help us to maintain the distinction, for they always speak of grid references and map references, which are just means of helping us to locate and speak about reality, the objects on the ground. Most other quantities that we meet in theoretical physics, such as energy and momentum, or electromagnetic field vectors or tensors, are like grid references, to enable us to identify a physical state and to make calculations about it. Theoretical objects such as fields, potentials, coordinates, electrons are not directly observable, we can observe only their experimental consequences, such as the acceleration of some particle or the pressure upon some surface, and so the question arises whether the theoretical objects have a physical reality. From one point of view, they have, for just as we may define the intensity of a light source by an operational procedure, equally we may, though by a less direct scheme, set up an operational definition of a potential or of the electron. An alternative approach is to say that quantities such as potentials or coordinates are solely objects of theory, but they are none the less defined operationally by the theoretical procedures into which they enter. Some quantities, such as electrons, may change their status from theoretical to observed quantities as the nature of the operational definition changes. In summary, all physical quantities are defined operationally, some with more observational content, others with more theoretical content. That might seem to blur any distinction between observed and theoretical quantities, but a clear distinction does remain in the purposes for which various quantities are employed. Observational quantities are employed to represent the empirical content of our interactions with the physical world, theoretical quantities to represent our abstract models of those interactions. They are not identical.

It seems reasonable to think that theory will correspond in some ways to the natural world but that in other respects it may reflect how we make observations. That there must be some correspondence seems clear. If we want to talk about objects in ordinary space of three dimensions, we must use quantities that correspond to the dimensionality of the space, essentially triplets of numbers, vectors in some conceptual space which is also of three dimensions or in some way equivalent.

One view of theoretical physics is that it is no more than an instrument that enables us to calculate the results of observations, as just a calculating device or black box into which hypotheses are put and out of which come results of observations. Theory is indeed an instrument and it is justified by satisfactory calculations of observations; the interesting question is whether a successful theory is only an instrument, or whether it is more than an instrument and offers some insight into the way the world is independently of us. It seems likely that the interior of the black box should correspond in some way to the deep elementary structure of nature. Some consideration of those issues arose in connection with the discussion of why mathematics works and the implications of the theory of groups and their representations and they will be taken up again in the final section of this chapter. The distinctions between the natural world and our observations of it, and between features of theoretical physics as means of calculation and as true representations of the natural world, will be kept to the fore.

Operational definitions and the highly abstract nature of much theoretical physics, suggest at first sight that physics is more abstract than biology, but it can be argued on the other side that biologists interfere more with the subject in their observations, often killing it or destroying fundamental order, so that they depend at least as much as physicists on theory for the design, operation and interpretation of their observations.

8.3 The indefiniteness of nature

It is a matter of observation that observations cannot be repeated exactly however carefully the circumstances are matched. In part that is because the world of real observations is more complex than any model of it. We attempt to control the conditions as closely as we can but so many, so inappreciable and so obscure are the factors that affect any observation that it is impossible to ensure exact repetition. We hope to treat the variations in results as random noise and derive best values and estimates of uncertainty on the assumption that a definite physical process with a definite result underlies the actual observations.

Even on the assumption of an underlying regular predictable physical process, the result of any observation cannot be predicted with certainty, nor can any completely rigorous conclusion be drawn from the result – for example, a set of data might indicate a linear relation between two variables, but there will always be some chance that the relation is more complex. All physical argument is therefore at bottom probable, not certain.

It is sometimes possible by the careful design of an experiment and by great care in its execution to approach quite closely to the ideal experiment which does exactly what it is intended to do, free of adventitious phenomena. There are however physical phenomena which are inherently irreproducible in themselves, phenomena that we call chaotic. No matter how great the care to eliminate disturbing influences, the results of any one experiment will never agree exactly with another. Again, probable argument must be used, but now on account of the internal properties of the physical system itself and not outside disturbances.

Science is a communal activity, its validity established by agreement between those who practise it. Observations are counted valid when there is communal acceptance of them. Similarly, the results of probable arguments should be in a form that they can be accepted or rejected by the community of scientists, so that a theory of probability that is inherently subjective is unhelpful. The empirical basis of science consists of data that are the generally accepted results of generally accepted methods of observation, together with generally accepted estimates of their reliability. That empirical basis, like its theoretical interpretation, is not fixed once for all; it evolves as more powerful technology or wider experience or fuller understanding lead to rejection of some results that were previously accepted, or to acceptance of others that at first seemed inconsistent. The progress of observational science can be seen as increasing the extent of those phenomena of which the general acceptance is sustained through further experience.

Physics and theoretical physics as effective rational activities depend on the belief that despite the incoherence and chaotic nature of our actual observations of nature and of the behaviour of nature itself, there is an underlying consistency that makes it sensible to make logical connections, not between actual observations but between some idealised distillation or derivatives of them.

8.4 The coherence of nature

One of the arguments of this book is that we impose coherence on our observations of nature through the ways in which it is possible to make measurements. With such coherences established, it is then possible to set up correspondences with mathematical structures, often through the concept of representations of groups. In that way we can construct mathematical theories that reproduce the results of observation. Some complications arise from inherent uncertainties in our observations of nature and in nature itself, but they do not mean that we have to abandon the attempt to give a rational account of our observations, rather that we have to see how uncertain and chaotic behaviour must be consistent with and fit into very general principles of dynamics.

But is nature itself consistent? That brings us back to engineering. The distinctive feature of engineering is that we go beyond observations of a world that we do not alter, for in engineering we operate on the world. If we are to achieve effective results we have to proceed on the basis that actions will have the same effects when we build bridges or operate chemical plants as they do when testing materials or trying out reactions in the laboratory: we rely on the consistency of nature. The consistency to which we appeal is like that of the property of a group, that there should be definite rules of combination for actions in nature, just as we require for members of an abstract group.

Different arguments have been made to justify special relativity as the local geometry of physics. It is the geometry that leaves observations of distant events unchanged (Chapter 3). It is the appropriate geometry and electrodynamics to use in devising models of the internal structure of atoms and molecules to give correspondence with their observed properties (Chapter 4). In both instances we are applying special relativity to models. Beyond that there are applications to engineering. The design and operation of a synchrotron for the acceleration of charged particles with relativistic velocities have to be based on relativistic, not non-relativistic electrodynamics. Is not that telling us that is how nature is, and not just our observations of it?

It is certainly the case that in important ways the possibilities or impossibilities of measurement do determine the form of theory, but there is another side to that. We can only establish a frequency standard in the way that we do because nature provides us with regular sets of energy levels of atoms. Similarly if there were no electromagnetic radiation we could not observe events at a distance as we do and so we should not have special relativity in the form that we do.

I conclude that nature is consistent, that if it were not we could not organise the results of observation in the form of rational theories, but that in important ways the form of a theory is determined by the sorts of observation that the properties of the natural world allow us to make upon it.

We have seen that there are two ways in which nature appears to be consistent. In the first, it allows us to make observations and perform operations that are the basis for material representations of groups. In the second, the behaviour of any complex system exhibits features that are properties of the formal arrangement of interconnections rather than of the physics or chemistry or biology of constituents. Thus there is developing a substantial body of theory of wide application independent of the individual nature of a complex system. I do not suggest that those two major forms of order exhaust the coherence of nature, but they are the ones most obvious in the present state of the development of science and through them, as in the other ways discussed in this book, the observations we can make upon the natural world and the nature of that world itself, determine the forms that theories must take.

8.5 Back to a real world?

The principal arguments of this book have been, first, that observations are the data generally accepted by the community of physicists, secondly that the credence to be given to them is that generally accorded by physicists, and, thirdly, that there are aspects of theory which are determined by the nature of observations as they are. The first requirement of theory is that it should give a rational account, based upon the least possible number of assumptions, of what the community of physicists generally accepts as the results of observations. We do not go behind the ostensive character of the observations, we do not in the first instance ask that theory should inform us about the real world behind the observations. At the same time, it might be thought that such self-denying attitudes to theory (instrumentalist) and observation (operational) cannot be the whole story, for it seems that they may not give an adequate account of the success of physics in providing the basis for changing the world by engineering.

I consider that the instrumentalist view of theory is an inadequate account of the actual content and use of theory. The instrumentalist view in an extreme form asserts that theory is fundamentally a set of relations between observations, denies the need for theoretical quantities and argues that it should be possible to replace theoretical connections by direct relations between observations (Hempel, 1965). There has been considerable discussion of how theoretical objects are to be defined (see Sahlin, 1990 for an account of the topic and an exposition of the problems that F. P. Ramsey confronted) and I argued above that they are in fact defined operationally. I do not consider that the definition of theoretical objects is the essence of the matter and I argued in Chapter 6 that if a theory is to be successful, its structure should correspond to that of a set of observations, in particular when they are both representations of the same abstract group. The objects of the theory are then defined operationally by their place in the theoretical representation and the correspondence between theoretical and observed objects is established through the common group structure. The instrumentalist view of theory is an incomplete account of actual theory because it disregards the fact that all successful theories derive many consequences from few hypotheses and an adequate theory of theory must explain why that procedure is so effective. A theory of theory must also explain why a theory constructed from a few theoretical concepts can predict the outcome of observations that were not envisaged before the theory had been developed. I argued in Chapter 7 that successful prediction over an unforeseen set of observations implies a structure of the physical world that is independent of the ways in which we choose to observe that world.

The foregoing argument should not be taken to imply that there is only one theoretical structure that corresponds to a given set of observations. We know, in fact, that a theory that represents some observations may have to be modified to include others – the obvious example is the extension of Galilean relativity to special relativity. The earlier theory remains valid within its proper field but just as the new observations are an extension to the initial set of observations, so the earlier theory is a sub-group of the later one. Thus a successful correspondence between theory and observation will always be an incomplete statement of the representation of observation by theory, one that further observation and study may extend and improve.

The body of empirical knowledge in physics rationally integrated as theory is analogous to a literary text, though written by no single author but by the whole community of physicists up to the present time. Whatever may have been the influences upon the members of the communal author and the considerations that guided them, the text as it is today is free standing and autonomous. It must be read for what it says and not for how it was written when we, its readers use it to lead us to new observations in physics or to change the world through engineering.

We do not fully know and cannot now enter into the mind of Monteverdi when he wrote 'Il Ritorno di Ulisse in Patria', or of Shakespeare writing 'Twelfth Night', or Mozart composing 'Le Nozze di Figaro', yet we still respond to those texts with laughter or tears because in deep ways they presuppose much the same psychology that is ours today – self-important persons in absurd situations make us laugh, constancy repaid after many years moves us to tears. So it is with our physics text. Every time that a prediction such as the existence of the neutrino is verified, every time a bridge is built, a space craft sent to a planet, a very high energy accelerator operated, a transgenic mouse bred, the combination of observation and theory is confirmed and furthermore, an assumption behind the text is confirmed, that there is an extended domain in which it applies that is consistent with the domain in which it was first established. Prediction in physics, successful engineering, prove existence theorems that there are domains independent of the communal authorship of the present text of physics in which the conclusions of that text are valid.

The analogy with a literary or musical text suggests that the argument should not be pushed too far. It could be said that the reason that we respond as we do to Monteverdi or Shakespeare or Mozart is that our psychology has been formed, in part at least, by just those texts. At the same time there must be a common basis of feeling and rationality on which they could work. In physics, likewise, it might be maintained that engineering works because the text selects the domain of engineering. Two propositions must be distinguished here. The first is that we select phenomena that we can place in a logical scheme and for which prediction is successful, and call them physics. The second is that there is no physical world, independent of us as observers of it. The first proposition is reasonable but in so far as it is true, it does not entail the second. The reverse is rather the case. The assertion that we select what we are able to study rationally from a wider range of phenomena would be meaningless unless that wider range existed. The selection argument implies a real independent world.

The debate about what we can know of a world goes back in its modern form to Locke and Hume. Hume maintained that we can only perceive events and can have no necessary knowledge of causes, and that is equivalent to saying that an operational definition of observations together with an instrumentalist view of theory precludes any knowledge of an underlying physical world. (According to some current interpretations of Hume, he was in effect a realist who saw that causal knowledge is not secure – rather like a physicist who is a realist but not a naïve realist.) Hume's contention can be avoided by the use of inverse probability, as was seen in Chapter 7, and although the justification of induction has always been a notorious problem, yet empirically, as Ramsey argued (Sahlin, 1990) induction is convincing because it is successful. Successful physics and engineering cannot prove with complete certainty that a real independent physical world exists but they are consistent with it and physicists and engineers have no difficulty in acting as if it were so. Nothing has falsified, in the sense of Popper, the hypothesis of a real independent world.

The main purpose of this book has been to analyse procedures in physics but the arguments are not confined to physics and raise the question whether biological science is the same sort of intellectual activity as physical science and whether the forms and purposes of theory are the same for it. The large part played by theory in observation in biology has already been mentioned, as well as the lack of geometrical order and the importance of kinematical order. The different types of order in biology arise essentially from the fact that biology is dominated by dynamical processes driven by a continuous flow of energy through an organism – the nearest analogues in physics are control systems. Biology and physics have the same aim at bottom, to establish intellectual structures that enable us to put the results of observation into a logical system, but because the phenomena are different, we may expect different forms of theory.

There is a subjectivism to science, not of individual scientists but of scientists as a community (d'Espagnat, 1989) which affects the choice of observations and the acceptance of the results of observation. The argument of this book has been that observations made in those terms do determine in certain ways the form of theory while beyond that, successful prediction shows that they also incorporate and reveal properties of a natural world independent of the existence of scientists.

APPENDIX

1 The definition of time

The unit of time is defined by Resolution No. 1 of the 13th. International Conference of Weights and Measures (Conférence Genérale des Poids et Mesures (CGPM), 1967:

The second is the duration of 9 192 631 periods of the radiation corresponding to the transition between the two hyperfine levels of the ground state of the caesium-133 atom.

The scale of International Atomic Time (TAI) is realised in practice by a number of frequency standards at different points on the Earth and therefore in different gravitational potentials, with the corresponding relative differences of frequency amounting to nearly 2×10^{-13} at a height of 1700 m. Standards in geostationary satellites will be subject to greater differences. The most precise comparisons of frequency and time must take those relativistic shifts into account and the Consultative Committee on the Definition of the Second of the International Bureau of Weights and Measures (CCDS, BIPM) declared:

TAI is a coordinate time scale defined in a geocentric reference frame with the SI second as realised on the rotating geoid as the scale unit.

The phrase, *as realised on the rotating geoid*, means that all frequencies are to be reduced to values at sea level on the rotating

Earth, for the geoid is the surface of constant gravitational and rotational potential which, over the oceans, coincides with the mean sea level surface.

2 Astronomical reference systems

The XXIst General Assembly of the International Astronomical Union, by its Resolution A4, adopted the following Recommendations of its Working Group on Reference Systems.

Recommendation I

considering

that it is appropriate to define several systems of space-time coordinates within the framework of the General Theory of Relativity.

recommends

that the four space-time coordinates ($x^0 = ct$, x^1, x^2, x^3) be selected in such a way that in each coordinate system centred at the barycentre of any ensemble of masses, the squared interval ds^2 be expressed with the minimum degree of approximation in the form:

$$ds^2 = -c^2 d\tau$$
$$= -\left(1-\frac{2U}{c^2}\right)(dx^0)^2 + \left(1+\frac{2U}{c^2}\right)[(dx^1)^2+(dx^2)^2+(dx^3)^2].$$

where c is the velocity of light, τ is the proper time and U is the sum of the gravitational potentials of the above mentioned ensemble of masses and of the tidal potential generated by bodies external to the ensemble, the latter potential vanishing at the barycentre.

Recommendation II

considering

(a) the need to define a barycentric coordinate system with spatial origin at the centre of mass of the solar system and a geocentric coordinate system with spatial origin at the centre of mass of the Earth, and the desirability of defining analogous coordinate systems for other planets and the Moon,

(b) that the coordinate systems should be related to the best realisations of reference systems in space and time, and
(c) that the same physical units should be used in all coordinate systems,

recommends that

1. the space coordinate grids with origins at the solar system barycentre and at the centre of mass of the Earth show no global rotation with respect to set of distant extragalactic objects,
2. that time coordinates be derived from a time scale realised by atomic clocks operating on the Earth,
3. the basic physical units of space-time in all coordinate systems be the second of the International System of Units (SI) for proper time and the SI metre for proper length, connected to the SI second by the value of the velocity of light $c = 299\,792\,458\ \mathrm{ms^{-1}}$.

REFERENCES

Chapter 1

Cook, A. H. 1961. Precise measurements of the density of mercury at 20°C. II. Content method. *Philos. Trans. Roy. Soc.* A, **254**, 125–54

1975. The absolute measurement of volume (B. le Neindre and B. Vodar, eds.) *Experimental Thermodynamics 11, Experimental Thermodynamics of Non-reactive Fluids*, pp. 303–20. London: Butterworth

1977. Standards of measurement and the structure of physical knowledge. *Contemp. Phys.* **18**, 393–409

1992. Metrology and the logical structure of physics. In (L. Crovini and T. J. Quinn, eds.) *Metrology at the Frontiers of Physics and Technology. Proc. Int. School of Physics 'Enrico Fermi' Course CX*, 27 June–7 July 1989, pp. 99–111. Amsterdam, etc.: North Holland

d'Espagnat, B. 1989. *Reality and the Physicist*. Cambridge: Cambridge University Press

Eddington, A. S. 1953. *Fundamental Theory*. Cambridge: Cambridge University Press

Franklin, A. 1989. The epistemology of experiment. In (D. Gooding, T. Pinch and S. Schaffer, eds.) *The Uses of Experiment, Studies in the Natural Sciences*, pp. 437–60. Cambridge: Cambridge University Press

Gooding, D., Pinch, T. and Schaffer, S. 1989. *The Uses of Experiment*,

Studies in the Natural Sciences, pp. xvii and 481. Cambridge: Cambridge University Press

Hacking, I. 1983. *Representing and Intervening*. Cambridge: Cambridge University Press

Hesse, M. B. 1974. *The Structure of Scientific Inference*. Berkeley: California University Press

Jeffreys, H. and Wrinch, D. 1921. *Phil. Mag.*, **42**, 369–90

Kartaschoff, P. 1978. *Frequency and Time*, London, etc.: Academic Press

Pedersen, O. 1993. *Early Physics and Astronomy* (rev. edn.). Cambridge: Cambridge University Press

Petley, B. W. 1985. *The Fundamental Physical Constants and the Frontier of Measurement*. Bristol: Adam Hilger

Pickering, A. 1989. Living in the material world. In (D. Gooding, T. Pinch and S. Schaffer, eds.) *The Uses of Experiment*, *Studies in the Natural Sciences*, pp. 275–97. Cambridge: Cambridge University Press

Toraldo di Francia, G. 1981. *The Investigation of the Physical World* (Engl. trans.). Cambridge: Cambridge University Press

Ziman, J. 1978. *Reliable Knowledge*. Cambridge: Cambridge University Press

Chapter 2

Fröhlich, H. 1967. Microscopic derivation of the equations of hydrodynamics. *Physica*, **34**, 215–26

1973. The connection between macro- and microphysics. *Riv. del Nuovo Cimento*, **3**, 490–534

Ozorio de Almeida, A. M. 1988. *Hamiltonian Systems, Chaos and Quantization*, pp. ix and 238. Cambridge: Cambridge University Press

Petley, B. W. 1985. *The Fundamental Physical Constants and the Frontier of Measurement*. Bristol: Adam Hilger

Chapter 3

Anderson, J. D., Esposito, P. B., Martin, W., Thornton, C. L. and Muhleman, D. O. 1975. Experimental test of general relativity using time-delay data from *Mariner* 6 and *Mariner* 7. *Astrophys. J.*, **200**, 221–33

Chen, Y. T. and Cook, A. H. 1993. *Gravitational Experiments in the Laboratory*. Cambridge: Cambridge University Press

Fomalmont, E. B. and Sramek, R. A. 1977. The deflection of radio waves by the Sun. *Comm. Astrophys.* **7**, 19–33

Landau, L. D. and Lifschitz, E. M. 1971. *The Classical Theory of Fields, Course of Theoretical Physics* Vol 2, 3rd. ed. Oxford, etc.: Pergamon

Petley, B. W. 1985. *The Fundamental Physical Constants and the Frontier of Measurement.* Bristol: Adam Hilger

Reasenberg, R. D., Shapiro, I. I., McNeil, P. E., Goldstein, R. B., Breidenthal, J. C., Brenkle, J. P., Cain, D. L., Kaufman, T. M., Kemarek, T. A. and Zygielbaum, A. I. 1979. *Viking* relativity experiment: verification of signal retardation by solar gravity. *Astrophys. J.* **234**, L219–21

Shapiro, I. I. 1980. Experimental tests of the general theory of relativity. In (A. Held, ed.), *General Relativity and Gravitation.* New York: Plenum Press

Shapiro, I. I., Pettengill, G. H., Ash, M. B., Ingills, R. P., Campbell, D. P. and Dyce, R. B. 1972. Mercury's perihelion advance – determination by radar. *Phys. Rev. Lett.*, **28**, 1594–7

Toraldo di Francia, G. 1981. *The Investigation of the Physical World* (Engl. trans.). Cambridge: Cambridge University Press

Chapter 4

Berestetskii, V. B., Lifschitz, E. M. and Pitaevskii, L. P. 1982. *Quantum Electrodynamics.* Oxford, etc.: Pergamon

Cook, A. H. 1988a. A Hamiltonian with linear kinetic energy for systems of many bodies. *Proc. R. Soc. Lond.* A, **415**, 35–59

Penrose, R. and Rindler, W. 1984, 1986. *Spinors and Space-time*, vols. 1 and 2. Cambridge: Cambridge University Press

Chapter 5

Cook, A. H. 1988b. *The motion of the Moon.* Bristol and Philadelphia: Hilger

Jeffreys, H. 1933. The function of cyclones in the general circulation. *Pro-verb. Assoc. Météorologie int.*, Pt. 2. 219–33. Jeffreys, *Collected Works*, **5**. (1976) 257–69

Lindhard, L. 1954. On the properties of a gas of charged particles. *Kgl. Dan. Vidensk. Selsk. Mat. Phys. Medd.*, **28**, No. 8

Poincaré, H. 1910. *Science and Method.* Paris

1916. *Oeuvres complets*, **1**. Paris

Waldram, J. R. 1985. *The theory of thermodynamics.* Cambridge: Cambridge University Press

Whittaker, E. T. 1927. *A Treatise on the Analytical Dynamics of*

Particles and Rigid Bodies, 3rd ed. Cambridge: Cambridge University Press

Chapter 6

Fröhlich, H. 1977. Long-range coherence in biological systems. *Riv. del Nuovo Cimento.* **7**, 399–418
Jeffreys, H. 1973. *Theory of Probability.* Oxford: The Clarendon Press
Jeffreys, H. and Jeffreys, Bertha, S. 1972. *Methods of Mathematical Physics.* Cambridge: Cambridge University Press
Jeffreys, H. and Wrinch, D. 1921. *Phil. Mag.*, **42**, 369–90
Lloyd, C. E. R. 1991. *Methods and Problems in Greek Science.* Cambridge: Cambridge University Press.
Ziman, J. 1978. *Reliable knowledge.* Cambridge: Cambridge University Press

Chapter 7

Bernoulli, Daniel. 1760. An attempt at an analysis of the mortality caused by small pox and of the advantages of inoculation to prevent it. Translation in Bradley, 1971. *Small Pox Inoculation in 18th Century Medical Controversy*
Cook, A. H. 1990. Sir Harold Jeffreys. *Biogr. Mem. F.R.S.*, **36**, 301–33
de Finetti, B. 1974, 1975. *Theory of Probability*, vols. I and II. (Engl. trans.) London: Wiley
Jeffreys, H. 1973. *Theory of Probability.* Oxford: The Clarendon Press
Jeffreys, H. and Wrinch, D. 1921. *Phil. Mag.*, **42**, 369–90
Keynes, J. M. 1921. *Treatise on Probability.* London: Macmillan
Ramsey, F. P. 1926. *Truth and Probability.* (Published posthumously in R. B. Braithwaite (ed.) 1931.) *The Foundations of Mathematics and other Logical Essays.* London: Routledge and Kegan Paul
Sahlin, N-E. 1990. *The Philosophy of F. P. Ramsey.* Cambridge: Cambridge University Press

Chapter 8

d'Espagnat, B. 1989. *Reality and the Physicist.* Cambridge: Cambridge University Press
Hempel, C. G. 1965. *Aspects of Scientific Explanation.* New York: Free Press
Sahlin, N-E. 1990. *The Philosophy of F. P. Ramsey.* Cambridge: Cambridge University Press

INDEX

aberration, 43
action principle, 79
Ampère, 136
  unit, 28
Anderson, J. D., 50, 157
angle measurement, 36, 37
angular momentum, 58, 64, 65
  orbital, 25
  spin, 25, 58
annuities, 123
area, as tensor, 48
astrophysics, 14
atmosphere, circulation, 77
atomic beam, 16
  deflection, 17
atomic physics, 28
  standard, xi
  structure, 52, 53, 58
  transition, 59, 60
  wavefunction, 105
attraction, basin, 81
attractor, strange, 81
automorphism, 105

Baconian method, 13
base line, 36
Bayes, 123
Bayesian probability, 129
belief, degree of, 123, 128
  justified, 123
Berestetskii, V. B., 67, 158
Bernoulli, D., 125, 159
biological systems, 75, 119
biology, xi, 14, 70, 71, 100, 120, 141, 145, 151
Boltzmann distribution, 87
boost, 65
Born–Oppenheimer principle, 85, 88
Bose–Einstein statistics, 93
bosons, 93
bound states, 53
bound systems, 78, 86, 94
boundary conditions, 75
bridges, 136
Bureau International des Poids et Mesures (BIPM), 28

caesium-133, 9, 16, 19, 29
  frequency standard, 17
celestial mechanics, 82, 84, 88
chance, 123
chaos, xi, 12, 13, 72, 134, 146
charge reversal, 109
chemistry, 141
Chen, Y.T., 50, 51, 157
classical mechanics, 22
Clifford algebra, 57
collective properties, 90, 94
combinatorics, 124
communal practice of science, 146
commutation, 102
condensed matter, 82
Conféference genérale des Poids et Mesures (CGPM), 9, 153
confidence interval, 123
conservation, particles, 56

constants
  defining, 26, 29
  fundamental, 26
  operational definition, 11
  physical, 8
Cook, A. H., 5, 7, 50, 51, 66, 85, 125, 156, 157, 158, 159
coordinates, 40, 143
  Cartesian, 40, 103
  transformation of, 42
Copernicus, 99
cosine rule, 38, 39
creation, 100
crystals, 87
current, unit of, 28
cyclones, 78

d'Alembert, 125
deduction, 123
de Finetti, B., 124, 125, 127, 133, 159
degeneracy, 107
deism, 100
Delaunay, 115
d'Espagnat, B., 2, 3, 151, 156, 159
detector, 21
diatomic molecule, 101
dimensions of representation, 65
Dirac, P. A. M., 68
DNA, 142
Doppler shift, 42, 43
dynamics, 110
  chaotic, 116, 135
  classical, 70
  geometrical representation, 75
  Hamiltonian, 113, 117
  initial conditions, 71
  large systems, 90
  non-linear, 70, 71, 90
  variables, 71

Earth, rotation, 10, 24, 78
earthquakes, 100
economics, 14, 70, 71
economic systems, 119
eddies, in oceans, 78
Eddington, A. S., 1, 156
eigenfunction, rotation, 62
eigenstate, 142
eigenvalue, 18
Einstein, A., 33, 49, 50
electric field, 31, 32
electromagnetic forces, 143
electromagnetic tensor, 48
electromagnetic theory, 99, 136
electron, 59
electronic charge, 8
emission, stimulated, 10
energy, 91
energy-momentum
  operator, 54, 57
  vector, 33, 44, 45, 110
engineering, 7, 137, 147, 148
entropy, 93
enzymes, 120
ephemeris second, 10
epicycle, 99
epistemology, 1, 2, 135
equations
  of motion, 7, 24, 111
  time evolution, 18, 142
equivalence principle, 50
ergodic hypothesis, 91, 92
errors
  experimental, 122
  random, 122, 123
  systematic, 5, 122
Euclidean geometry, 35
  metrology, 35
evolution in time, 18, 19, 55, 58
exact solutions, 114

Faraday, M., 136
Fermi, 99, 136, 137
Fermi–Dirac statistics, 93
fermion, 93
fiducial probability, 123
field theory, 113
filter, magnetic, 9, 10
Fisher, R. A., 123, 129
fixed point solution, 118
flow, in dynamics, 70
fluid dynamics, 72, 77
Fomalmont, E. B., 49, 157
Fourier components
  of standard frequency, 11, 20
  phase, 41
four-vector, 41, 143
  acceleration, 44
  potential, 48
  velocity, 43
frame, distant, 40
Franklin, A., 5, 156
frequency, 41, 42
frequency standard, xi, 8, 9, 16ff., 23, 142, 147
  radio transmission, 10, 27
Fröhlich, 119, 157, 158

general relativity, *see* relativity, general
genes, 120
geology, xi
geometrical representation, of dynamics, 75
geometry, Euclidean, 35
geophysics, 14
Gibbs, W., 143
God, 100

Gooding, D., 2, 156
gravitational potential, 51
gravitational waves, 99
Greek science, 100
grid reference, 143
groups, 102
  Abelian, 102
  applications, 68
  continuous, 102, 103, 109
  finite, 102, 108
  general linear (GL), 73, 105
  point, 58, 101, 108
  rotation, 61, 65, 103, 104
  space, 88, 108
group representations, 146
  bases, 60, 103
  character, 107
  dimension, 107
  irreducible, 106
  physical, x, 60, 103
  theory, 67, 103

Hacking, I., 4, 157
Hall effect, quantum, 8, 27, 28, 46
Hall resistance, 27
Hamiltonian, 18, 22, 23, 56, 58, 111, 112, 114
Hamiltonian equations 25
Hamilton–Jacobi equation, 25
harmonies of nature, 100
heliocentric system, 131
Hempel, C. G., 149, 159
Henry's law, 98
Hertz, 99
Hesse, M., 3, 157
Hume, D., 151
hydrogen maser, 10
hyperfine states, 9, 10, 13, 18

identity operator, 21
induction, 13, 123, 151
inference, 1
  scientific, 130
initial conditions, 71
inoculation, smallpox, 125
instrumentalism, of theory, 4, 6
International Astronomical Union (IAU), 8, 11, 50
inverse probability, 128, 130, 151

Jeffreys, B., 104, 159
Jeffreys, H., 15, 77, 98, 104, 117, 124, 125, 126, 129, 131, 133, 135, 138, 157, 158, 159
Josephson effect, 27, 28, 46, 48

Kartaschoff, P., 10, 157
Kepler, J., 99
Keynes, J. M., 124, 125, 127, 133, 159
kilogramme, 8
Klein–Gordon equation, 55

Lagrangian, 99, 111
Lagrangian points, 85
Landau, L. D., 33, 157
Laplace, 123
least action, 117
Leibnitz, 131
length, standard, 26
lepton, 59
Lie group, 63, 110, 111, 112, 113, 115, 117
life assurance, 123
Lifschitz, E. M., 33, 157, 158
light, gravitational, deflection, 49
likelihood, 123, 129, 130
Lindhard, L., 89, 158
linear physics, 116
linear theory, 12
liquids, 82
literary text, 149
literary theory, 3
Lloyd, G. E. R., 100, 159
Locke, 151
logistic equation, 73, 77
Lorentz group, 63, 67, 110
  transformation, 30, 33, 41, 54, 71, 109
lunar theory, 85

magnetic field, 31
magnetic moment, intrinsic, 58
manifold, 75
many-body dynamics, 67
map reference, 143
mappings, 137
mass
  rest, 49
  standard, 8, 28, 48
mathematics
  nature and structure, 97, 100
  objects of, 5
  and physics, x, xi
  representing observation, 98
  in theory, 12
  uses, 97
matrices, 101, 103
Maxwell, J. C., 30, 99, 136
  equations, 30, 99
  distribution, 92, 93
measurement
  and physics, x, 13
  precision, 7
  standards, 7, 142
mean value, 122
mechanics
  classical, 22
  quantum, 23
metals, 82, 89

meteorology, 72
mètre, 8
metric, 49
  coefficients, 36
  Schwarzschild, 50
metrology, Euclidean, 35
microphysics, 143
models
  linear, 134
  non-linear, 134
molecular structure, 59
molecular vibrations, 101, 103
Monteverdi, 150
Moon, motion, 82
motion, equations, 7
  integrals, 80
Mozart, 150
muon, 59

natural world, 140
nature, consistency, 137, 138, 147
networks, 119
neutrino, 136, 150
Newton, I., 7, 83, 99
Noether's theorem, 79
noetic domain, 3
non-linear phenomena, xi, 70, 71
normal modes, 86, 87, 94, 101, 103, 105, 109
Nozze di figaro, Le, 150
nuclear magnetic resonance, 141
null vector, 42
numerical solutions, 119

observables, 40, 51, 143
observation
  astronomical, 141
  high energy physics, 141
  of world, 140
observations
  elements of, 4
  operational definition, 5, 6, 91, 141, 145, 148
  ostensive character, 148
  theoretical content, 3, 4
Oersted, 136
oil, exploration for, 100
operational definition, 94, 145
  of frequency standard, 11, 19
  of theoretical quantities, 144
operators, 21, 103, 142
orbital theory, 114, 115
orthogonal functions, 113
Ozorio de Almeida, A. M., 25, 157

Pauli matrix, 57, 66
pattern recognition, 99, 120
particle accelerator, 147
Pedersen, O., 12, 157
Penrose, R., 57, 158
perfect gas, 77, 78
perturbation theory, 73, 85, 114, 131
Petley, B., 9, 26, 157, 158
phase space, 75
phonons, 87, 109
physics
  consistency, 132
  constants of, 8
  logical structures, xi
  subjective elements, 3
physical knowledge, 1
physical world
  structure, 100
  stability, 135
Pickering, A., 2, 4, 157
Pinch, T., 2, 156, 157
Planck's constant, 8
planetary motion, 99
Plato, 6
plasmas, 82, 89
Poincaré, H., 70, 71, 72, 73, 80, 158
Poincaré group, 109, 110
point group, 101
Poisson bracket, 112, 113
Popper, K., 151
population dynamics, 120
potential, electric, 31
power, standard of, 28
prediction, xi, 1, 13, 99, 100, 116, 135, 138, 147
pressure, 93
Principia Mathematica, 126
probability
  axioms, 126
  Bayesian, 124, 127
  frequency definition, 124
  inverse, 128, 130
  joint, 127
  operational definition, 133
  objective, 133
  ordering of, 125
  prior, 129, 130
  subjective, 123, 133
probable argument, x, 14, 146
Ptolemaic cosmology, 99, 131
Pythagoras, 12, 100
  theorem, 36

quantum Hall effect, *see* Hall effect, quantum
quantum mechanics, xi, 2, 102, 111, 142
  formal structure, 20
  many-body, 84
  time evolution equation, 29, 55, 58
quantum standards, 9

radiation, black-body, 4
Ramsey, F. P., 124, 125, 149, 151, 159

random variations, 145
ray paths, 36
realism, 2
real world, 148
Reasenberg, R. D., 50, 158
reflexion operation, 103
relativity
  general, 11, 35, 36, 49, 50, 51, 99
  special, 11, 33, 50, 79, 107, 142, 147
representation theory, 67, 103
resistance, to theory, 4
rest energy, 110
Ritorno di Ulisse in Patria, 150
rotation group, 61, 65, 103, 104
  eigenfunctions of, 62
Russell, B., 126

Sahlin, N-E., 124, 151, 159
sampling, 123
Schaffer, S., 2, 156, 157
second, 8, 10
seismic reflexions, 100
selection of data, xi
separable equations, 114
Shakespeare, 150
Shapiro, I. I., 49, 158
simplicity principle, 131
social construction of science, 3
social nature of science, 14, 151
social studies, 14
sociology, 3
Solar system, 85
soliton, 116
space-craft, 150
space-time interval, 109
special relativity, *see* relativity, special
speed of light, 8, 33, 34, 36, 51
  near Sun, 50
spherical harmonics, 63
spin–orbit interaction, 58
spinors, 66
Sramek, R. A., 49, 157
standard deviation, 128
standards, 7
  atomic, xi, 16–19
  electrical, 7, 8
  length, 7, 26
  mass, 7, 8
  operational definition, 19
  quantum, 9, 16–20
  time and frequency, xi, 7, 9, 10, 16–19
statistical mechanics, 81
statistical models, 94
statistical thermodynamics, 123
statistics, 14, 122
sub-groups, 102
Sun, orbital motion, 24
superconductivity, 89
survey, electromagnetic, 35
symmetry, 58, 59, 60, 61, 99, 101
  broken, 108
  groups, 60
  operations, 60
  translational, 108
synchrotron radiation in biology, 141
Systéme Internationale des Unités, 8, 26

tangent operator, 63
technology, 137
tensor, electromagnetic, 48
theory
  correspondence with observation, 6, 39
  as instrument, 6, 144, 148
  as model, 6
  operational definition in, 124
  quantities of, 40, 143, 149
thermal energy, 89
thermodynamics, 80, 82
three-body problem, 12, 82, 83
time, ephemeris, 24
  evolution in, 18, 22, 23, 55
  sidereal, 24
time, scale of, 20, 34
  mechanical standard, 24
Toraldo di Francia, G., 4, 5, 40, 157
trajectory, chaotic, 92
  of solution, 74, 75
transformations, 101–105
transgenic animals, 142, 150
transit time, 36
two-body problem, 83
Twelfth Night, 150

uncertainty, 12

variables, dynamical, 24, 71
variance, 122
vector potential, 47
vectors, 103
voltage, as operator, 21
volume, measurement of, 5

Waldram, J., 81, 91, 158
wavefunction, 18, 19, 109, 142
wave vector, 41, 42
Weyl transform, 25
Whitehead, 126
Whittaker, E. T., 83, 158
Wittgenstein, 10
Wrinch, D., 15, 98, 129, 131, 138, 157, 159

Ziman, J., 3, 12, 96, 100, 157, 159

Printed in the United States
By Bookmasters